——— 世界文化遗产 ———
WORLD HERITAGE

HUIZHOU SCHOOL ARCHITECTURE OF CHINA

——— 樊炎冰 主编 ———
Chief Editor　Fan Yanbing

中国建筑工业出版社
CHINA ARCHITECTURE & BUILDING PRESS

联合国教科文组织

文化和自然遗产保护协定

联合国遗产委员会已将

皖南古村落西递及宏村

列入世界遗产名录

本名录确认：凡列入本名录的文化或自然场所，即具有特殊的和世界性的价值。为了全人类的利益，需要加以保护。

颁发日期：2000.12.2

UNITED NATIONS EDUCATIONAL,
SCIENTIFIC AND
CULTURAL ORGANIZATION

CONVENTION CONCERNING
THE PROTECTION OF THE WORLD
CULTURAL AND NATURAL
HERITAGE

The World Heritage Committee
has inscribed

*Ancient Villages in Southern Anhui-Xidi
and Hongcun*

on the World Heritage List

Inscription on this List confirms the exceptional
and universal value of a cultural or
natural site which requires protection for the benefit
of all humanity

DATE OF INSCRIPTION
2 December 2000

DIRECTOR-GENERAL
OF UNESCO

序言

罗哲文

在我国封建社会的后期，明、清两代王朝大约五六百年的时间里，曾经出现了"无徽不成镇"的商业经济繁荣的局面。其影响面之大，几乎波及全国。徽州人不仅推动了生产的发展，社会经济的繁荣，而且也促进了科学技术和文化艺术的繁荣与发展。他们之中，许多人为官致富或致富为官，他们同时还培育了许多科学技术和文化艺术的杰出人物，推动了科学技术、文化艺术和社会经济的繁荣与发展。

建筑，是政治、经济发展的具体表现，是科学技术与文化艺术的载体，徽州人在其显达和致富后，首先想到的就是修建宅第、祠堂、书院、牌坊……等等光宗耀祖的大业。由于建筑业的长期发展，工程技术、工艺操作技艺经验的成熟，逐渐形成了具有特殊地区风格的徽派建筑，成为我国古代建筑中，一个独特的流派，建筑艺术园地中一朵绚丽的奇葩。其中，不少已成了国家各级重点文物保护单位，有的还列入了世界文化遗产的名录，十分珍贵。

徽州古称新安，曾下辖黟县、歙县、休宁、婺源、祁门、绩溪六县。山灵水秀，人文荟萃。仅清代休宁一县中状元者就有13人之多，位居全国之首。其他县乡，或有"父子尚书"、"同胞翰林"，或有"一门八进士"、"一镇四状元"者，文采风流，莫此为甚。徽州历史文化源远流长，内容广博。新安程朱理学、皖派朴学、新安画派、新安医学、徽剧戏文等等，对徽派建筑的形成，都有着深刻的影响。

诚如本书所言，徽派建筑虽然多姿多彩，但从特定的角度上观察，人们举目一望，就可归纳为一系列显明的特点：外观上的粉墙黛瓦，四水归堂的天井庭院，错落有致的山墙造型（马头墙），气宇轩昂的门楼门罩，巧夺天工的雕刻技艺，高大宽敞的楼上厅堂，临水观赏的飞来椅、美人靠等等。

散落于黄山白岳之间，承负着数百年明清风霜的古民居、古祠堂、古牌坊，号称徽派建筑三绝。例如：具有"民间故宫"美誉的黟县宏村"承志堂"，占地面积数千平方米，气势宏大，包括内、外院，前、后堂，东、西厢，以及书房厅、小客厅、鱼塘厅，还有曲径通幽的"排山阁"与"吞云轩"。该处的楼阁厅堂典雅灵巧，其中砖、木、石雕精美异常，抬头望去，额梁上"百子闹元宵"、"唐肃宗宴百官"、"郭子仪做寿"、"九世同堂"等雕刻，金碧辉煌，一一映入眼帘，国内极其罕见。总而言之，宏村"承志堂"无疑是徽州古民居中的一处经典之作。

历史上，祠堂作为缅怀先祖、加强宗族内部凝聚力的一种特殊公共场所，在徽州地区星罗棋布，盛极一时，几乎达到了"无姓无祠，无宗无堂"的地步。现存绩溪龙川胡氏宗祠与歙县呈坎村罗东舒祠（又称宝纶阁）被当代古建筑保护专家称之为国家的两大祠堂。龙川胡氏宗祠，前后三进，建筑面积达1000多平方米，高大的门楼方梁上雕刻精美图案，一为"九狮滚球遍地锦"，一为"九龙戏珠满天星"。两旁木梁上雕刻"三国演义"各种人物场景，玲珑剔透，栩栩如生，可谓独步一时。呈坎村罗东舒祠四进四院，建筑面积达2000多平方米，包括照壁、棂星门、碑亭、仪门、两厢、露台、享堂、后寝，整个布局严谨，气势雄伟恢弘，令人叹为观止。

徽派建筑的村落选址依山傍水，大都与风水观念相关，而宅第厅堂的形态布局、图案雕刻，又暗含儒道佛哲学意蕴。例如棠樾保艾堂坐南朝北（与常例相反），且大小房间建有108间，这无疑同道教崇拜七星北斗以及向往三十六洞天、七十二福地的观念有关。棠樾村口七座牌坊前后贯穿一气，遥相呼应，取名题词极力张扬的忠孝节义，显然源自儒家。至于在墙壁、门窗、梁柱上随处可见的如意图案、莲花图案、文字图案，都可以说明佛教思想对徽派建筑的影响。其次，黟县宏村的牛形布局，西递村的船形布局，都借助当地具体的地形条件，随后加以人工开凿修饰，表面上似乎看来与古代流传已久的风水观念一一吻合，但从深层次观察，却与当代国内外流行的生态建筑学与景观建筑学有相通之处。所有这些方面前人研究不多，开拓不深。本书主编熟悉现代建筑设计、传统方术与儒道佛哲学，故能引经据典，从这几个层面来提示徽派建筑不同寻常的深层奥秘与文化底蕴。

中华文明渊源悠久，灿烂辉煌。倘如说，诞生于华夏大地的先秦诸子哲学、唐诗宋词是一大笔不朽的精神遗产，那么留存至今的江南园林、徽派建筑则是积淀了优秀传统文化的物质载体，值得后人大力研究。本书主编樊炎冰先生历时八年数十次下徽州，跋山涉水，考察"一府六县"的古民居、古祠堂、牌坊群以及典型村落布局形态，数历寒暑，批阅大量文献，终于撰写了图文并茂的《中国徽派建筑》一书，为徽派建筑研究作出了重大的贡献。本人为徽州古建筑的保护和西递、宏村申报世界文化遗产的工作等，曾多次到徽州进行考察，深受徽派建筑历史文化艺术之感染与吸引，与之结下了深厚的情缘，今见此书之即将出版，不仅乐见其成，并且冀望其早日问世，爰书数语以为序。

Preface

Luo Zhewen

In about five or six hundred years of the Ming and Qing Dynasties, the late period of China's feudal society, there was a time of prosperous commercial economy marked by the saying of "No Huizhou merchants, no towns and cities". The impact of Huizhou merchants could be felt almost all over the country. Among them, many became officials for seeking profits, or sought profits for holding the official posts. But at the same time they brought up a number of talented people in science and art, not only promoting the development of production and the prosperity of the social economy, but also accelerating the progress of science and technology, as well as art and culture.

Architecture reflects the development of politics and economy, and it is the physical carrier of science and technology, and art and culture. When the Huizhou merchants became rich or held high official posts, the first thing which came to their minds was to build residences, ancestral halls, academies, memorial archways, etc., in their home town for the sake of bringing honour to their ancestors. Based on the long development of building industry, the rich experience of building technology and the maturity of skills and practice, Huizhou architecture has formed its own special local characteristics and become a unique school in China's ancient constructions, an exotic flower in the garden of the architectural art. Among the precious buildings, many are listed as the most important historical sites to be given special protection by the central and local governments and some are officially inscribed on the World Cultural Heritage List by the UNESCO.

Huizhou was called Xin An in ancient time, governing 6 counties of Yixian, Shexian, Xiuning, Wuyuan, Qimen and Jixi. It is a place full of beautiful scenes and outstanding people. In Xiuning County alone, there were 13 Zhuangyuan (title conferred on the one who came first in the highest imperial examination) only in the Qing Dynasty. In other counties, there once appeared "father and son ministers", "brother Hanlins", or "eight Jinshi in one family" and "four Zhuangyuan in a small town". Huizhou had a long histry of culture and high reputation in many aspects of literature, philosophy, arts and crafts, such as Xin'an Neo-Confucianism headed by Cheng brothers and Zhu Xi, Anhui School of philosophy, Xin'an School of Painting, Xin'an School of medicine, Anhui Opera, etc. They all deeply influenced the formation of the Huizhou-school Architecture.

As is said in this book, the features of the Huizhou architecture can be summed up as follows, though various in shape and size: white walls and black tiles in appearance; courtyards and skywells to admit daylight and rain, meaning "receiving wealth"; high and steep and well arranged horse-head walls; magnificent gate towers and shieldings; exquisite and beautiful carvings; tall and spacious halls upstairs and enjoyable beauty chairs, etc.

The ancient folk residences, ancient ancestral halls and ancient archways, scattered between the Huangshan and Qiyun Mountains and existing for several hundred years during the Ming and Qing periods, are regarded as the three masterpieces of Huizhou-style constructions. Among them is the Chengzhi Hall of Hongcun in Yixian County, which has a local name of "folk imperial palace". It covers an area of several thousand square meters and contains the outer compound, inner compound, front hall, rear hall, east and west wing rooms, studies, sitting hall and fish-pond halls. Besides, there are a hall for mah-jong playing and a hall for opium-taking, which are quiet and secluded. All the halls and pavilions here are elegant and exquisite, and the wood, brick and stone carvings in them are delicate and lively, the best of which are the rare carvings of "A Hundred Children Celebrating the Lantern Festival", "Emperor Su Zong of Tang Dynasty Entertaining the Officials", "Guo Ziyi's Birthday Celebration" and "Nine Generations Living Together" on the beams and architraves. In a word, Chengzhi Hall of Hongcun is undoubtedly the classical and representative work of the ancient folk dwelling houses in Huizhou.

Ancestral halls, as special public places of worshiping ancestors and strengthening the cohesion of a clan, were all the vogue in history and spread in large numbers all over the Huizhou District. It seems that almost every family clan had its own ancestral hall. The extant Ancestral Hall of the Hu Clan in Longchuan of Jixi County and Baolun Pavilion of Chengkan Village in Shexian County are listed as the two ancestral halls of national value by the present experts of ancient architecture protection. The Ancestral Hall of the Hu Clan in Longchuan covers an area of more than 1000 square meters with three sections. The surface of the square beams under the eaves of the huge gate tower are decorated with delicate carvings: one of them is the pattern of nine lions and the other is that of nine dragons. The wooden beams on both sides are carved with figures and stories from The Three Kingdoms, which is very special and exquisite. Baolun

Pavilion of Chengkan Village has four sections and four courtyards, covering an area of more than 2000 square meters and including the screen wall, Lingxin Gate, stele pavilions, ceremonial gate, corridors, platform, Xiangtang (reception hall) and Qindian (ceremonial hall). The whole hall is well arranged and in an imposing manner, acclaimed by people as the acme of perfection.

Hamlets in Huizhou are usually situated either at the foot of mountains or by the riverside, mostly relating to the concept of Fengshui. However, the plan and shape of residences and halls, as well as the patterns of carvings, embodies the philosophical connotation of Confucianism, Taoism and Buddhism. For example, Bao'ai Hall of Tangyue faces the north, which is contrary to the convention, and has 108 rooms, big or small, which is related to the concept of Taoism. The seven archways in front of Tangyue Village are arranged in a line and named according to the teachings of Confucianism. The patterns of carvings on the walls, doors, windows, beams and columns, such as Ruyi (an S-shaped ornamental object, usu. made of jade, formerly a symbol of good luck), lotus and characters, reflect the impact of Buddhism on the Huizhou constructions. Then, the ox-shape planning of Hongcun of Yixian and the boat-shape planning of Xidi, taking the advantage of the practical local terrains and adding some artificial decorations, are in accordance with the traditional concept of Fengshui on surface, but in depth these plans are in harmony with the current eco-architecture and landscape-architecture which are very popular both at home and abroad. Few people are involved in the study of these fields and there are not many findings. The editor of this book, familiar with modern architecture designs, traditional divination and philosophy of Confucianism, Taoism and Buddhism, can therefore search, by citing various theories, for the unusual secrets in depth and cultural meaning of the Huizhou architecture from different aspects.

China has a long and splendid civilization. If we say that the philosophies during the pre-Qin period, emerged in China, and poetry during the Tang and Song Dynasties are our immortal mental heritage, the extant gardens in south Yangtze and the Huizhou constructions are the physical carriers of our excellent traditional culture, reserving a thorough study and research. Mr. Fan Yanbing, chief editor of this book, had paid tens of visits to Huizhou in eight years, traveling afar under difficult conditions and investigating the ancient folk residences, ancient ancestral halls, groups of archways and typical hamlets. After reading large volumes of records, he finished the writing of this book, which is a great contribution to the study of the Huizhou architecture. Since I was engaged in the reservation of the Huizhou ancient buildings and the application of Xidi and Hongcun for the list of world cultural heritage, I went to Huizhou many times and have been deeply moved and attracted by the historical and cultural arts of Huizhou-school architecture, thus being irrevocably committed to it. I am happy to see the publication of this book, hence the preface.

徽派建筑漫记

田木相

一

还没有去过徽州之前，我就神游在徽州的葱茏的峰峦和清淑的山水之间！

记得读汤显祖文集时，他对徽州，对徽州山水的向往，特别是他对于徽州的建筑名胜的描绘，就曾让我萌生了对徽州的魂系梦绕之情。汤显祖曾因去徽州未成而写下了以下的诗句："欲识金银气，多从黄白游。一生痴绝处，无梦到徽州。"（《吴序怜余乏绝，劝为黄山白岳之游，不果》）写下了他对徽州的一片痴情。晚年，他终于实现了他的梦想，到了徽州的海阳（休宁），写下了著名的《坐隐乩笔记》，这是一篇海阳镇的山水园林的赞美诗：

> 予尝闻海阳之地，松萝奇秀，不让匡庐、九嶷、巫峡，心窃慕之。戊申秋，偕陈子伯书裹粮履杖其间。海阳里旧门墙之士，复彬彬在侧。果飞障茵葱，列（嵷）回合。入其里，曰高士里；望其庐，曰坐隐先生宅也。门人皆知先生者，交口而述先生。先生诗文之外好为乐府、传奇种种，为余鉴赏。正与余同调者。余亟欲阐扬之。先生有园一区，堂曰环翠，楼曰百鹤，湖曰昌湖。其中芝房茵阁露榭风亭，传记大备，诸名贤之诗歌辞赋不可指数。先生灌花浇竹之暇，参释味玄，雅好静坐。间为局戏，黑白相对，每有仙着，近成《订谱》（即《坐隐先生精订捷径棋谱》）行于世。人号坐隐先生。盖先生屏却世氛，独证妙道，有日月在手，造化生心之意。其精神常与纯阳通，提醒假寐，仍见于乩仙语沥沥。缘字丹书，独于先生泄其秘义，称为全一真人者，信不诬也。先生行无辙迹，言无瑕谪，夫岂自见自矜，亦岂炫奇骇世哉！盖不必觌霞标，接玄尘，雅知其为通籍于八公，藏名于三岛者也。予奇其事而爱慕之不已，故为先生记。（万历戊申秋九月临川汤显祖为友人汪其朝先生记）

这里对海阳的赞美，也是对于整个徽州的赞美，其山松萝奇秀，其园环翠百鹤，其阁榭风亭犹如仙境……这般绮丽的风光怎不令人神往！

我还没有去过徽州，我就倾倒在徽州古老的独树一帜的文化氛围之中！就感受着这片充满神奇的土地的神秘色彩！

从读清人曹文埴的诗《咏西递》，就知道西递是一个为人赞不绝口的"小桃源"，甚至把徽州真的当作陶渊明所写的《桃花源记》的世外桃源之所在地了。这更增添了徽州的神奇和神秘感。

> 青山云外深，
> 白屋烟中出。
> 双溪左右环，
> 群木高下密。
> 曲径如弯弓，
> 连墙若比栉。
> 自如桃花来，
> 墟落此第一。
> ……

这首诗，也让我梦系徽州。西递，西递！是多么诱人去探索的去处！

总之，所有这些，都让我未识徽州真面目，已是徽州梦里人了！

二

我不迷信，但是，我却相信命运。在我看来，命运就是一种坚韧地自我追求、痴迷的梦想同客观条件碰撞的结果。它似乎很神秘，而细细地追味似乎也很合乎情理。

说来都很神。新世纪之初，我早期的一个学生——本书的主编樊炎冰，突然找上门来，非邀我为这部画册写一篇介绍性的文字，邀我亲自作一次徽州之行。你所梦想的，所盼望的，突然降临，自然我很爽快地答应下来。我说：我一定去，我一定写，但是我不一定写得好！

3月，正是春花烂漫的季节，炎冰陪我开始了徽州之游。出屯溪，先是西北行，去歙县、黟县，这是徽州民居保留最多最好的地方。出城不久，在眼前展开的是一片小平原，路边的水杉、三角枫和冬青树缀满着春天的气息，而满眼的菜花，更把田野山峦装饰得一片金黄！不久，即进入齐云山境，汽车在山谷中穿行，迂回崎岖，山峦起伏，确有进入世外桃源之感。

炎冰显然是有意这样安排的，第一个让我看的景点就是被联合国列为世界文化遗产的宏村。这是一个保存比较完好的古村落。它真的把我震呆了，我真的不敢想象，在这样僻远的深山中，居然有着这样一个天人合一的佳境，一时间，恍惚疑非人间。那如镜月塘和南湖，那依山面水犹如一只卧牛般的村落，那富丽的承志堂，让我流连忘返。再请我看的就是西递村了，这个如同船形的村落，同宏村一样，也被联合国列为"世界文化遗产"。

炎冰又带我去棠樾看著名的牌坊群，天淅淅沥沥地下起雨来。在村口，七座牌坊由东北向西南依次按忠、孝、节、义逶迤展开：鲍象贤尚书坊、鲍逢昌孝子坊、鲍文渊继妻节孝坊、乐善好施坊、鲍文龄妻节孝坊、慈孝里坊和鲍灿孝行坊，每一座牌坊都有它的故事。在一座贞节牌坊前，我凝望着那"节劲三冬"和"脉存一线"的题词，似乎看到这贞节牌坊的女主人公凄苦的身影，听到了她们啼哭之声。此刻，雨，依然戚戚地下着！

三

徽州，这不仅是一个地理的概念，一个历史的概念，一个文化的概念，而且它是中国古老文化的象征。因此，徽州成为一门专门的学问——徽州学。

先说说徽州的地理。

徽州位于安徽南部的黄山和齐云山之间。

据说，大禹时代，在徽州地域，就栖息着彪悍勇猛的土著先民，三民族和古越族。到汉代，史称"山越"。秦代开始设黟县和歙县。唐代设有歙州，下辖绩溪、歙县、休宁、黟县、祁门、婺源六县。

宋宣和三年，即公元1121年，歙州改称徽州，起名徽州是据所属绩溪有徽山、徽水、徽岭之故。北宋末至清代的徽州，其辖境相当于今安徽省的歙县、黟县、休宁、祁门、绩溪和江西省婺源等县，既当年的"一府六县"。

在祖国辽阔的大地上，徽州的地理环境极为特殊，山地和丘陵占据十分之九。徽州东有昱岭、大鄣山，西有浙岭，南有江滩，北有黄山，地理环境显得格外闭锁险峻，素有"山限壤隔，民不染他俗"之称。

它处于"山岭川谷崎岖之中"，位于"吴楚分源"之界。战国时代，吴楚争雄，徽州即处于两国中间。在婺源的浙源乡的浙岭上，至今还保留着"吴楚分源"的界碑。从历史上看，它避开多次的战祸，偏安一隅，成为一个少受战乱的地方，成为一个相对独立而安全的地域。它似乎又是格外的外向，徽州人走遍了祖国各地，成就了著名的徽商，也造就了大批的徽州官宦之家。因此，在徽州到处都可以看到皇家赐下的牌坊，即使在偏远的山村，也可以看到高门豪宅。

就是在这样一个独特的地理环境中，形成它特有的历史轨迹。到过徽州的人，无不感受着它曾经拥有的巨大的财富，无处不昭示它的辉煌；也无不体验到它深厚的历史文化积淀，一碑一桥，一屋一物，一草一木都烙印着绚丽的历史文化的记忆。在历史上，它不但是理学的桑梓之邦，而且是中国京剧文化的故乡。可谓物华天宝，人杰地灵。在徽州，诸多领域都曾涌现过杰出的代表人物，理学大师朱熹，集中体现了徽州儒学的深厚根基和贡献，而著名的反理学家戴震更标志着徽州活跃的哲学思维。这里，还曾出现过著名的明代的珠算大师程大位，以及著名的画家渐江、汪采白、黄宾虹等。至于文人学士的著作更是不可胜数，据统计，明清两代的经、史、子、集的著述，总计为2486部，蔚然壮观。

四

我生在北方，对于北方的民居，自然十分熟悉。无论是东北的大院落，还是北京的四合院，还是山西的多进的大院，基本是以平房，或砖，或砖坯结合，或土坯作为建筑材料的。而徽州的建筑，则以石材、木料为主体，加之它的地理形势，直接影响着徽派建筑的风格和特色。"居庐之制，因居山国，木植价廉，取材宏大，坚固耐久，今元代所营之室，村之旧者犹数见焉，明代建筑不足奇矣。然以山多田少，病居室之占地，多作重楼峻垣。"（民国《歙县志·风土》

卷一）这些，直接影响着徽派建筑的风格和特色。

对于徽派建筑的风格特色，很多专家都曾作出很好的描述和艺术概括：

"朴素淡雅的建筑色调，别具一格的山墙造型，紧凑通融的天井庭院，奇巧多变的梁架结构，精致优美的雕刻装饰，古朴雅致的室内陈设等。"(《徽商研究》）

"就徽州同类建筑而言，同是民居，均有明敞透亮的天井，高峻腾飞、跌落有致的马头墙，昂然挺立、逐层跳出的斗拱，形象凸现、生机盎然的三雕（木雕、砖雕、石雕）。"(《徽派建筑艺术》）

的确，那富有节奏感和韵律感的马头墙；那高高的院墙内，狭窄的天井所形成的狭窄的空间，都给我留下了深刻的印象。但是，它最初给我的感受是更为独特更为深刻的：当你第一次跨进这里高大的院门，却进入一个逼促的阴暗的、潮湿的甚至还带有一点霉味的空间。站在那个狭窄的天井中，仰天望去，眼界中天空被割成一个小小长方形，人，真的成为井底之蛙了。我不曾感到特别明敞，也许，正赶上阴天下雨的缘故。即使是放晴的时候，我想这个天井也不如北方的院落那么敞亮。人住的房间，也是比较狭窄的。到了楼上，内眷居住的地方，光线就更为灰暗；那些空守闺房的少妇，成年累月凭依着窗栏所能看到的不过就是眼前那灰色的高高的墙，还有头顶上的一小片天空。身处其中，就感到《徽州女人》的故事是真实的。

只有当我从直接的氛围感受进入对这些民居的沉静的观赏之中，才使我从逼促的感觉中走出来，慢慢地体味到徽派建筑的特有的美的风格特色。特别是作为历史文物，作为观赏的美学对象，徽派建筑的价值是具有其不可忽视的历史价值和美学意义的。也许随着历史的流失，它的建筑艺术的持久力和文化的魅力越加显现出来。

马头墙是徽州民居最具特色的构件，每间住宅的两侧，都有高出屋面的山墙，它沿着屋顶的斜坡而逐次分层跌落，而且一律都涂上粉白色。引人注目的是封火墙的端部都有造型，武官人家都砌成黛青色的马头形，文官人家都砌成黛青色的官印型。特别是马头墙的造型，给人以动感，好像骏马奔腾。从远处观望徽州的村落，错落的马头墙，形成一种万马腾空的景象。

马头墙的色彩基调也是徽州民居色彩的基调，它以灰白色作为基调，又辅以青色，所谓粉墙青瓦。从远处望去，灰白色的村落闪现在一片葱茏的绿树丛中，又有绿水青山的映衬，显得它格外的和平、安祥、宁静。徽州民居的色彩美学，独具一格，简单、素朴的色彩，却有了最为耐人寻味的欣赏价值。的确如有学者所说的："特别是几百年后的今天，经过长期的日晒、风吹、雨淋，墙面上的白粉早已斑斑驳驳地脱落，从而出现一种冷暖相交的多次复色。尽管它失去了白色的明朗、单纯，却因此产生了一种厚重的历史感。"（余治淮：《桃花源里人家》）

天井也是徽州民居的一大特色，它建于门堂之间。敞开的大堂与大门之间，多了这样一个天井，它似院非院，在高高的院墙的闭锁之中，又带来一片湛蓝的天空，一个相对开阔的空间，显然使之更为敞亮明朗，空气也更为流通清新。据说，原是为了防盗而建的高大的墙体，为人增添了安全感，但是，却又带来采光、通风和心理压迫诸多不便，于是才又设计了"天井"。倒是一位日本建筑学家对徽州的天井具有独到的审视眼光，他说：

"在中国的住宅建筑中经常有的'院子'，在徽州却被二层楼的建筑物所框围住，作为'光庭'而被室内化了。据说它是明亮敞开的天空之井，故称之为'天井'。楼下中央，正对着大门朝着天井开放的部分被称为'正堂'或'祖堂'。其他的房间被壁板等遮盖住，无法知道其中情况。除去四周关闭着的房间，在这个四周被高墙所围的'天之井'的底部所展开的十字形空间，完全是一个在从石块和砖头的表面所透出的微微冷气中，在紫烟飘逸的丝丝微风中，在透过窗上的木雕而摇曳着的清冷光线中，在与外界完全隔绝的静寂之中，与中国的激烈变动似乎完全无关地沉睡至今的空间；又是一个带着一种难以动摇地把某种中国文化紧密封存的厚重的空间。在这种'天井'里，有着一种无论是在日本还是西欧的住宅里我至今从未见过的透明而静谧的光线。在此之后，我们被引导去参观的所有民房里都有着'天井'。虽然在所有的二层建筑的底部都有着这样一个静谧的空间，但是它们各自的造型千姿百态，使用方式也是多种多样的。我深深地为'天井'这种奇妙的建筑空间所魅惑。在这里我似乎觉得我能看到徽州民房建筑的魅力所在，进而看到江南文化的特点。"（日本建筑思潮研究所：《住宅建筑》）

在一种比较建筑文化的视界中，徽州民居的"天井"的哲学意蕴和美学内涵，被这位日本学者细微而深刻地捕捉到了。

再有就是徽州民居的斗栱，堪称一绝。所谓斗栱，是斗和栱的合称。斗为方形坐斗，栱为弓形肘木。"方形之斗，弓形之栱，相互勾连，交错层叠，形成一体，谓之斗栱。斗栱位于梁柱之间，起着关键的支撑作用，又起着中介过渡作用。"徽州的斗栱，吸取了明清两代斗栱建筑艺术的精华，并加以创新，形成其独具的特色："既富于繁缛斑斓之美，又富于简明朗丽之美"。"它的简，是指部件单纯，组合明朗，线条清晰，则偏重于简，它以简驭繁，寓繁于简。特别是，它以绚丽多姿、五彩缤纷的藻饰涂抹造型，使它那简洁的构架形式又增添了繁缛的色泽与光辉"（参见《徽派建筑艺术》）。我站在房檐下，仔细观赏这些藏在屋檐下的千姿百态的斗栱形象，有时产生一种不可思议的疑惑和情思：我怀疑把斗栱做得这样精美，是否是必要的，特别是这不易为人发现的地方；但是，我又不能不佩服徽人的美感，是那么细腻，那么柔韧，真可谓无微不至、无孔不入了。哪怕在这屋檐之下，也让斗栱这样一个建筑部件成为一件艺术品。

五

徽派建筑的另一特色就是它的三雕了，即木雕、石雕和砖雕。徽派建筑同徽雕水乳交融，密不可分。有徽州建筑之处，必有徽雕。无论是民居、祠堂、庙宇、牌坊、楼塔……处处都可看到精美的徽雕艺术。也无论是梁、柱、枋、斗栱、门、墙，徽雕都融入其中，成为这些部件有机构成的部分，徽雕把建筑部件的实用性同审美性结合起来。

徽雕分三种：

一是木雕，进得院内，满眼扑来的就是十分精细的木雕作品，在梁柱上，在斗栱上，在雀替、窗槛、家具、屏风上，以及文房用具上，无处不展示着徽州木雕的风采。形成所谓雀替雕饰、斗栱雕饰、额枋雕饰、钩挂贴饰、轩顶雕饰、脊檩包袱、元宝雕饰、楼沿雕饰、月梁雕饰等，构成家庭的浓郁的文化氛围和优雅的审美环境。人居其中，俯仰之间，顾盼左右，那精美的木雕作品即展现在你的眼前，给你以视觉的愉悦，性情的陶冶。

我也见过一些木雕，但是却没有徽州木雕给我如此深刻的印象，特别是那些系列木雕形象，如草船借箭、群英会、汾阳府（插图1）等木雕群像，还有系列木雕故事，如三国故事，甘露寺、长坂坡、三战吕布等，这些都可以说是徽雕的绝活。它的镂刻艺术精细灵巧，剔透玲珑。如西递承志堂的《唐肃宗宴官图》，高一尺，宽六尺，横向展开的五六张桌席上，大约有三十余个身姿神态各异的人物，或弹琴，或下棋，或绘画，或书法，竟然有一个人正在掏耳朵，颇富情趣。其构图也十分完整，其雕刻精巧已达到炉火纯青的境界。在约五六厘米的木板上，雕刻出七八个层次，给人以立体感。

本书所拍的木雕，有的堪称绝作，如余庆堂的"九狮雀替"（插图2），高1.2米，其构图奇妙，雄狮形象跃然。其他如胡氏宗祠的木雕、俞氏宗祠的木雕、豸峰堂的"双凤朝阳"木雕，都是难得一见的木雕杰作。

其次是砖雕，它用特制的青砖雕刻出来，诸如门罩、窗罩，特别是砖雕的门楼（插图3），以精美的砖雕来装饰建筑的外部空间，那是徽州的一绝。砖虽是水磨青砖，质地细腻，但却没有木质的韧性、石质的硬度，仍然显得粗些；可是它在徽州人手中，却化为艺术。我在北岸村，亲眼目睹了门罩上的砖雕，它那严谨的构图、凸显的图案，以及在其中央雕刻出的园林景色和人物群像，其手工之细腻精致，有的竟然有九个层次，较之木雕还多，真令人叹为观止。至于砖门楼，那就更为讲究，什么垂花门楼，什么字匾式门楼，牌楼等等，庄重典雅、威严气派。

再次是石雕。徽州民居也多用石雕，如抱鼓石狮、门墙、栅栏、台阶、基座、匾额等。特别是庙宇、祠堂、古碑、牌坊、桥梁、宝塔等，更是施展石雕之美的地方。在查济的二甲祠，我看到精美的基座、石鼓等。

在黟县，到处都是石雕艺术品。我特别感兴趣的是这些石雕作品，往往蕴藏着主人的精神寄托。据说西递西园的主人胡文照曾是一位具有远大抱负的官员，他把姜子牙渭水垂钓的故事勒石刻于门楣之上，或许寄托着他辞官归隐之意。

"漏窗"，本用以调节空气，便于采光，但在徽州人手中，也化为艺术。如"松石"漏窗、"竹梅"漏窗，"树叶"漏窗，"琴棋书画"漏窗等，或组成"岁寒三友"，显现着主人的超逸清高之旨趣；或以树叶暗示"落叶归根"（插图4）的情感；或以"喜鹊登梅"的图案，寄寓对幸福生活的期冀；或以"鲤鱼跳龙门"（插图5）的图案，憧憬美好的未来。这些，从一个侧面透视着徽州建筑的文化品格。

作为石雕之一的碑刻，也许不为人所重视，但是它却有着不可忽视的历史价值和艺术价值。在徽州，无论是在田野间，村落前，小溪边，抑或是院内、屋旁、桥畔，处处都可以看到这些大大小小的碑刻。或是历史沧桑的记录，或者人物的小传，或是工程的记录，它成为一种珍贵的历史资料。同时，它也是艺术品，是书法艺术的珍藏，是诗文的宝库。像宏村南湖书院所保留的《春日过雷冈怀王心览》，就是明代的一块诗碑。而保存在黟县文物局的二通《石雕双松梅竹图》，就是就是诗画交融、珠联璧合的绝妙的艺术珍品。

也许人们不禁要问，为什么徽州建筑在院内和室内的装饰上如此下功夫，把心思全部用于"内装修"上，从门窗设计，到梁柱雕饰，可谓"无孔不入""无微不至"？据记载，在明代即有对于民间建筑的种种限制和规定，不准在建筑规模上越规，不准使用金碧辉煌的装饰，等等，违反了甚至有杀头之祸。于是徽商就在规定的范围内，在自家的院内自家的屋里，极尽其能事；用心于住宅的布局结构，使其更加紧凑坚固，精心地装修使其秀丽精美。这样，不仅是其将来告老还乡颐养天年的住所，也可与乡党争阔斗富。于是密集型劳动为代价的砖、木、石三雕艺术便在这里应运而兴，由此形成了徽州建筑特有的装饰规模和风格。

六

徽派建筑之所以具有魅力,固然建筑的自身有着它的特色和风格,但是,你不得不为之倾倒的是它的建筑群体所构成的村落,以及村落同自然的和谐所形成的浑然天成、天人合一的境界。

为了更好地了解这些村落的整体构思和整体规划,先说说徽州村落的"聚族而居"的特点。

"聚族成村到处同"(《徽歙竹枝词》),聚族而居是徽州古代村落的最突出的特征。可以说,徽州是家族制度极为盛行的地区之一。大约汉代以后,就形成了以宗族血缘关系为纽带,代代相传繁衍生息而成的同族聚居的村落。在《寄园寄所寄》中指出:"千年之家,不动一丕;千丁之族,未尝散处;千载谱系,丝毫不紊;主仆之严,数十年不改"。据历史记载,聚族而居,始于"中原衣冠"入徽。大批的北方人口聚族南迁,进入徽州之后,往往便选择一些山谷隘口、险要易守之地落脚,聚族而建村落,如此繁衍。如黟县的西递村,就是胡姓宗族聚居之地。据《胡氏宗谱》记载,胡氏祖先姓李,唐朝皇帝昭宗后代。西递,为胡氏五世祖胡士良所发现,至今将近三百年,子孙繁衍了数十代。聚族而居,为一个村落的整体规划提供了可能。

到明代中叶,徽州的村落已经形成了独特的形态和景观。到清代中叶之后,就发展到一个全盛时期。不但民居建构模式达到一个成熟完美的程度,而且就整个村落的整体规划也有着一个和谐、合理的整体布局,其他如牌坊、祠堂、桥梁、园林、池塘、书院等文化民俗建筑设施,也融于村落的整体环境之中。

历经数百年,至今徽州仍然保留着一些完好的村落,如黟县的西递、宏村、南屏、关麓,歙县的呈坎、许村、棠樾,婺源的汪口,绩溪的瀛洲等。这些村落已经成为徽派建筑的展览馆、山水园林的游览地和历史文物的博物馆。

聚族而居,取人与人和;择山选水,取人与自然和。乾隆《汪氏义门世谱·东岸家谱序》中说:"自古贤人之迁,必相其阴阳向背,察其山川形势。"于是山谷、峰峦、隘口,以及水口、溪水、渡口、河流,就成为徽州人建村之胸襟,布局之关注,美学之把握。

正如《徽派建筑艺术》中所概括的:"或依山而居,扼山麓、山坞、山隘之咽喉;或傍水而住,据握河曲、渡口、汊流之要冲。村落星罗棋布,重重叠叠;建筑纵横交错,密密层层。可谓组合有序,多样统一,形态各异,气象万千。然其布局,却有规律可循。若据高俯视,从宏观上把握,则不外有:呈牛角形者,如婺源西坑民居;呈弓形者,如婺源太白司民居;呈带状者,如婺源高砂民居;呈之字形者,如婺源之梅林民居;呈波浪形者,如黟县西递民居;呈牛胃形者,如黟县宏村民居;呈星云团聚形者,如歙县潜口民居;呈蜿蜒曲折形者,如歙县江村;此外,尚有呈半月形者,呈方印形者,呈弧形者,呈直线形者,呈点状形者,等等。"

黟县的宏村就是一个典型。宏村始建于南宋。最初,不过只有13间房,为汪氏家族聚居之地。据说,汪氏数代不兴,以为没有善用此地风水,特请来著名的风水先生休宁县的何可达来观察地势,他走遍了宏村的远山近水,查清了山脉和河流的走势,确认宏村的地势犹如一只卧牛,于是按照牛形来进行宏村的整体规划。在中国,以一个村落为建设的整体进行规划的,还不多见。这正是徽州建筑的一大特色。在上百年间,先是利用村中一自然泉水,建成一个半月形的池塘,作为"牛胃",之后,又开凿了一条长达四百米的水渠,作为"牛肠",后又在村西虞山溪上,建起四座木桥,作为"牛脚",于是便形成了"山为牛头,树为角,屋为牛身,桥为脚"的牛形村落。最后,在村南将百亩良田开掘出一个小小的人工湖泊——南湖,成为另一个"牛胃"。这样就使宏村成为湖光山色,家家流水,怡然自得,诗意盎然的环境。我虽然走过不少的村落,南方的北方的,但是如宏村这样让我流连忘返的世外桃源般的巧夺天工的仙境,只有它了。我为祖先的诗性智能所折服,我为前人的天人合一的哲学所倾倒。

正因聚族而居,使徽州地区的祠堂建筑格外突出,几乎是村村有祠堂。宗族为血缘而连接起来的群体,在漫长的中国历史中,始终是中国社会组织的基石,也是宗法观念的基因。对祖先的膜拜,对宗族的信仰,形成了宗祠建筑,同时也形成宗祠的精神,宗祠的情结。它深刻地渗透在徽派建筑的格局和建筑品种与建筑风格之中,它既是一个显现的建筑体系,也是一个潜隐的精神力量。

徽派的建筑,举凡民居、牌坊、古塔、古亭、书院、园林、直到祠堂,都是环绕着宗祠而展开和布局的。

宗祠,集中体现了对于先祖的膜拜,使它自然成为祭祀祖先的殿堂。如宝纶阁,珍藏着历代皇帝赐予呈坎村罗氏宗族的诰命、诏书等恩旨纶音,就典型地传达了宗祠祖先膜拜的意识和观念;而建于祠堂之前,或者是村落之前的牌坊,显然也在于宣示宗族之显赫的社会地位,光辉之业绩。塔、阁、亭等,同样是展示宗族之道统、威严、信仰和恩德的标志。由宗祠而构成徽派的建筑系统。但无疑这些建筑中所凝结的无声音乐自然也是宗法的封建的。

以西递村为例,据记载,其鼎盛时期,胡氏家族的祠堂就有34座。有宗祠、总宗祠、分宗祠和家祠。有人形容说西递的祠堂,可谓祠堂林立,宛如一个"祠堂世界"。其富丽之程度,其耗资之巨大,令人瞠目结舌。据本始堂造址遗存《乐输建造宗祠》石碑刻记:"耗资白银六千九百四十两,均为族内各房头祠会和富户乐输。"由此,也可见宗法观念之深了。

七

为什么在这样一个山峦起伏层叠的封闭的环境中,有着这么多具有徽州特色的建筑,这样的美好的村落,有时真是叫人百思不得其解的。追索起来,其原因甚多,但是,徽商的崛起是徽州建筑文化繁荣的经济基础。

徽商系指以乡族关系为纽带所结成的徽州商人群体。大体从明代中叶兴起,直到清朝末年衰落。据《歙县志》记载:"彼时盐业集中淮扬,全国金融几可操纵,致富较易,故多以此起家,席丰礼厚,闾里相望,……令其所遗仅有残敝之建筑,可想见昔年闳丽而骄惰之习"。徽商主要经营的是盐业和金融业,他们致富之后,即回故里"修祠堂,建园第"。

徽商建屋,不惜钱财,极尽富丽堂皇之能事。据《重订〈潭滨杂志〉序》记载:"潭滨者,黄氏所居村名,亦谓之潭渡……至前明而簪绂特盛,及国朝而益炽,文献之迹,详于往牒矣。若夫风俗之粹美,室庐之精丽,皆他族所罕俪。兹编所记,虽若琐屑,然承平丰乐之景象可想见也。乾隆以后,故家巨室亦稍稍替矣。然旧德犹在,闾井晏然,以崇儿时所见,犹然一乡望族也。"可见当年潭渡建筑是何等的壮观,何等的气派!

《野议》(宋应星著)对徽商为什么会将大量的商业利润投入建筑,说得更为清楚:"商之有本者,大抵属秦、晋与徽郡三方之人。万历盛时,资本在广陵者不啻三千万两,每年子息可生九百万两。只以百万输帑,而以三百万充无妄费,公私具足,波及僧、道、丐、佣、桥梁、楼宇。尚余五百万，各商肥家润身，使之不尽，而用之不竭，至今可想见其盛也。"而叶显恩在其《明清徽州农村社会佃仆制》中指出："由于官府的庇护和享有豁免税收等特权而取得优惠利润的徽商，是一般商人所不能与之竞争的。他们并没有感到有改为经营商品生产的必要……所以，当商业资本超过经营商业所需要的数量之后，超过部分便如上所述，或挥金如土地耗费在'肥家润身上'……"这些道出了徽州建筑兴盛的原因。

"无徽不成镇"。据《徽州府志》（1699 年刊行）记载："徽州富民，尽家于仪、扬、苏、松、淮安、芜湖、杭、湖诸郡，以及江西之南昌、湖广之汉口，远如北京，亦复挈其家属而去。"

而徽商建屋，富丽堂皇。描写当年徽商之盛，清人诗《屯浦》：

> 一片遥帆势若奔，客舟来集若云屯。
> 将归巧趁秋风便，欲落仍衔夕照痕。
> 山外人烟迷翠霭，渡头沽舶聚黄昏。
> 喧闹晚市明灯火，不是江南黄叶村。

徽商在外经营，当回归故乡建设自己的家园时，也将外面世界的建筑艺术带回去。譬如苏州、扬州的园林建筑盛名天下，徽商将苏州、扬州的喜建园林的爱好也带到自己的家乡，兴起营造园林之风：歙西溪南有果园，许承尧《歙事闲谭》录有："琐琐娘，艳珠也，妙音声。明嘉靖中，新安多富室，而吴天行亦以财雄于丰溪，所居广园林，侈台榭，充玩好声色于中。琐琐娘名聘焉，后房女以百数，而琐琐娘独殊，姿性尤慧，因获专房宠。时号天行为百妾主人，主人亦名其园为'果园'。"

八

徽商的财富自然是造就辉煌的徽州建筑文化原因之一，但是，徽州建筑的深厚的文化内涵，更展示了徽派建筑的深远的博大的文化渊源。

有的学者指出：徽州古村落有着深厚的人文关怀，它主要表现在以下几点：一是"以山水养人，怡情励志"；二是"以文教育人"；三是"以礼仪聚人"。（贺为才：《徽州古村落的人文关怀》，《光明日报》2001 年 7 月 19 日。）

的确，徽派建筑深蓄着儒家文化的底蕴，或者说透露着浓郁的儒雅之风。吴敬梓在其《儒林外史》中是这样描写扬州的徽派住宅的："当下走进了一个虎座的门楼，过了磨砖的天井，到了厅上。举头一看，中间悬着一个大匾，金字是"慎思堂"三字，两边金笺对联，写：'读书好，耕田好，学好便好；创业难，守成难，知难不难。'中间挂一轴倪云林的画。书案上摆着一大块不曾磨过的璞。十二张花梨椅子。左边放着六尺高的一座穿衣镜。从镜子后边走进去。两扇门开了，鹅卵石砌成的地，循着塘沿走，一路的朱红栏杆。走了进去，三间花厅，隔断中间悬着斑竹帘。"这同西递敬爱堂的格局，甚至对联都如出一辙。清代徽州哲学家戴震有言："虽有贾者，咸近士风"（《戴东原集》卷 12）徽州文人汪道昆说："夫贾为厚利，儒为名高。夫人毕事儒不效，则弛儒而张贾；既侧身飨其利矣，及为子孙计，宁弛贾而张儒。一弛一张，迭相为用。"（《太函集》卷 52）

这些描写自然是表层性的，但透过这些表层徽州却有着深厚的文化积淀和文化传统。徽州虽以"商贾之乡"驰名，但它也以文风昌盛之地而著称。远则不说，宋元以降，徽州就呈现出教育发达、人才辈出的特点。徽州虽系偏远的山区，但又成为人才聚居之地。其原因是在多次战乱中，大批的有钱之人，有才之人，有地位的人，纷纷到这里避乱，于是有"虽十家村落，亦有讽诵之声"的美誉。而徽商也有"贾而好儒"之风。徽人或"先贾后儒"，或"先儒后贾"，或"亦儒亦贾"。明清之际，徽州六县，讲学成风，单是书院就有54所，其中尤以紫阳书院名气最大。在此期间，这些书院为徽州培养出大批人才，据统计，明朝举人298人，清朝举人698人；明朝进士392人，清朝则有226人。一方面是徽商的儒化，也称缙绅化；一方面则是官僚化，据统计，在清代，单是歙县出现了大学士4人，尚书7人，侍郎21人，都察院都御史7人，内阁学士15人。胡适说："我们徽州人在文化上和教育上，每能得一个时代的风气之先……因此在中古以后，有些徽州学者，他们之所以能在中古学术界占据较高位置，都不是偶然的。"（《四十自述》）。

但是，也应看到传统的礼教对于徽州建筑的负面影响，所谓"美人靠"就是其中之一。按照礼教的约规，女孩子要住在楼上，而且是不能轻易下楼的，于是一些人家便在二楼客厅靠天井的一侧，设置一排靠椅，让她们观看楼下和外间的景物，这一排靠椅就称作"美人靠"。徽派建筑正是在这样一个深厚的文化传统中兴盛起来并自成一派的。

九

最后，不能不谈谈主编樊炎冰先生。

在外间看来，能够组织出版这样一部巨作的，必然是一个政府的部门，一个实力雄厚的科研单位，一个具有远见的出版机构。能够主编这样一部巨制，一定是一位十分精谙安徽民居的研究专家，一位十分有名望的学者。其实，都不是。樊炎冰先生，完全靠他对安徽民居的痴爱，对摄影的痴情；完全靠着他一个人的积累和精力，自己投资，自己和朋友一起调查，自己和朋友一起拍摄，完成这部著作的。因此，此书的成功，就是一个发生在开放改革时代的奇迹，是一个值得表彰的创举！

我听炎冰自己讲过，也听他的朋友讲过他的种种经历，讲过他十多年来为了完成这样一部著作所经历的种种艰辛。他说，"这部书是我的心血凝成的，是许许多多的熟悉的朋友和陌生的朋友的热情造就的。"

樊炎冰，1984年毕业于安徽建筑工程学院，这自然使他同建筑结成姻缘。但是，使他对安徽民居产生兴趣，还在于他分配到安徽省城乡规划设计研究院后一次偶然的工作机遇——参加规划院考察安徽民居的工作。他当时作为一个初来规划院的年轻人，除了听人说过安徽民居外，几乎对其毫无了解。但是在将近二十天的考察中，除去划归江西的婺源没有去，安徽所属的原徽州的五个县几乎都走遍了。这些徽派建筑，给了他十分强烈的震撼。他十分惊讶的是：安徽民居具有如此独特的建筑的美学的特色，拥有如此深厚的文化内涵，而且保存得如此完整。他被这样的灿烂的文化遗迹迷住了，由此注入他的心灵，融入他年轻的生命之中。

如他所说，他对安徽民居接触得越多，感情越深。而促使他要把安徽民居拍摄下来的念头，是他读了刘敦桢教授的《中国住宅概说》和张仲一等著的《徽州明代住宅》之后，他发现书上所提到的明代的安徽民居，再去寻找时已不复存在了。这就深深刺激了他。他痴呆地望着那些明代民居的废墟，那永远再也不能复现的遗迹，十分痛心。他眼看着这些民居被拆掉，拆下的门、窗、砖、梁……作为文物出卖，徽州的子孙在败坏着先祖的文化遗产……他痛心极了。1992年，他发誓要把现在还屹立在皖南土地上的这些珍贵的建筑拍摄下来，他要把它全部拍摄下来，保存一套完备的安徽民居的历史的形象档案。

从1992年开始，他人已到广东工作，但是却情系徽州，每年都数次地回到徽州。他深知要实现这样一个大的计划，不是他一个人所能完成的。于是他广交朋友，几乎徽州各县的文化局的领导，或者是当地的专家，都成为他的老师，他的朋友。他每次去，必然去拜访这些老朋友，虚心地向他们请教。这样，不但使他了解那些公开展览的民居，更使他深入到一些鲜为人知的深宅大院。

炎冰特别介绍他结交的一位徽州的朋友——许智勇先生。许先生对于徽州的民居十分熟悉，既懂方言，又会驾车。炎冰说："许智勇先生几乎陪同我跑了十年的徽州民居考察工作，对完成这本书的出版功不可没。"还有此书的另二位摄影师，一位是中国建筑工业出版社建筑摄影家张振光，一位是安徽摄影家吴广民，都是为樊炎冰的精神所感动，自愿地热情地投身这个计划的。为了此书，不辞辛苦，不遗余力。

炎冰的精神也感动着当地群众。在宏村，一位村民得知樊炎冰多次来宏村拍摄以及他的宏伟计划后，十分感动，就主动找到樊炎冰，为之提供一些还不为人知的宅子。保艾堂，据记载其原有规模甚为壮观，有房108间，但是，当他们去拍摄时发现已十分残缺败陋了。当业主得知樊炎冰的拍摄计划后，这位老者，就根据他的记忆，按照原来的样子，画出了108间的平面图。像这样的事例举不胜举。

炎冰全然凭着他的一片对祖国的文化遗产的痴情，一种发自内心的责任感，不但完成了一项对祖国对人民对历史极为有意义的事业，而且，也完成了他自身文化品格的塑造。在这样一个商品经济的大潮流中，在人文学术为世人冷落的社会空气中，他能把自己的财力、物力和人力投入祖国的文化积累的神圣事业之中，是十分难得的，我自己也是深受感动的。

当我结束这篇已经不算短的漫话时，我祝愿本书的主编樊炎冰先生，能够继续这种文化事业，也希望人们喜欢这部书。

2002年1月5日
于北京

Some Notes about Architecture of Huizhou School

Tian Benxiang

I

Before I went to Huizhou, I had long made spiritual tours around the green mountains and clear waters in Huizhou.

When I read the selected works of Tang Xianzu, the most famous playwright of the Ming Dynasty in China, I was deeply moved and attracted by his descriptions of the Huizhou architecture, his admiration of the beautiful scenery there. He once wrote a poem to express his regret at and longing for the trip to Huizhou when he failed to accept his friend's invitation of the travel:

"Want to know business and prosperity?

Go to Huangshan and Qiyun Mountains.

My lifetime infatuation is just left there,

For I have no chance of visiting Huizhou."

In his late years, he paid a visit to Haiyang (present Xiuning County) and his dream came true. He was so excited that he wrote a well-known travel note in praise of the superb landscape—the green mountains, clear waters, delicate gardens and wonderful buildings. This article impressed me so much that it fired my imagination.

Before I went to Huizhou, I had been overwhelmed with her unique cultural atmosphere and the mystery of this Shangri-La.

When I read the Song of Xi Di, a poem by Cao Wenzhi of the Qing Dynasty, I got to know this wonderland and I almost regarded the place as the haven of peace, described in Tao Yuanming's famous prose The Story of the Peach Blossom Valley.

Before I went to Huizhou, I had dreamed of it from dawn to dusk!

II

I don't believe in superstition, but I believe in fate. In my opinion, fate is something resulting from the complex of one's tenacious self-pursuit, persevering dreams and objective conditions. It seems to be mysterious, but actually it is reasonable if one ponders over it carefully.

Everything seems magic. At the very beginning of this new century, Mr. Fan Yanbing, one of my early students and the chief editor of this book, came to me unexpectedly and insisted on inviting me to a trip around Huizhou and asking me to write an introduction to this picture album. What you had dreamed of and longed for came to you all of a sudden. For sure I accepted his invitation readily.

In March of that year, the season when spring had come and flowers were in bloom, I began my trip to Huizhou accompanied by Mr. Fan. We left Tunxi and went northwestward to Shexian County and Yixian County, where the traditional folk houses of Huizhou are most and best preserved. Soon, we arrived at the Qiyun Mountains. The car went on the rugged path through valleys with undulating hills around and we really felt as if we had come to Shangri-La.

It was obvious that Yanbing had arranged all these on purpose. The first sight he showed to me was a well-preserved ancient hamlet—Hongcun Village, which had been listed on the World Cultural Heritage by the United Nations. I was really astonished at the sight. I could not imagine that in such a remote mountain area there was a place like this where heaven and man could be harmonized so well. At that very moment, I thought I was in paradise. I could not tear myself away from the beautiful scenes: the mirror-like Moon Pond and South Lake, the village at the foot of a hill and facing a stream in shape of a sleeping cow, and the imposing Chengzhi Hall. The next place we visited was Xidi Village. The shape of the village is like a boat and it's also listed on the World Cultural Heritage by the U.N.

When Yanbing took me to Tangyue to see the famous group of memorial archways, it began to rain pitter-patter. At the entrance to the village, seven archways stood from northeast to southwest in the order of the Chinese traditional virtues—loyalty, filial piety, chastity and friendship. Each archway had its own story. I was lost in front of the archways in the rain.

III

Huizhou is not only a geographic, but also a historic and cultural concept. It is one of the symbols of the Chinese ancient culture. Hence there is a special discipline of learning—Study of Huizhou.

Let's have a look at the geographic situation in Huizhou first.

Huizhou locates between the Huangshan and Qiyun Mountains in the south of Anhui Province.

It is said that during the reign of Emperor Yu (21st-16th century B.C.), many intrepid and valiant aborigines such as Sanmin and Guyue nationalities lived in Huizhou. In the Han Dynasty (206 B.C.—220 A.D.), this place was called "Shanyue". Yixian County and Shexian County were first set up in the Qin Dynasty (221-201 B.C.). In the Tang Dynasty (618-907 A.D.), Shezhou Prefecture was set up, governing 6 counties including Jixi, Shexian, Xiuning, Yixian, Qimen and Wuyuan.

In the year of 1121, that is, the third year of the reign of Xuanhe in the Song Dynasty, Shezhou Prefecture changed its name into Huizhou, for the reason that Jixi County, subordinated to Shezhou Prefecture, was famous for the Hui Mountains, Hui Rivers and Hui Ridges within its area. From the late Northern Song Dynasty to the Qing Dynasty, Shexian, Yixian, Xiuning, Qimen and Jixi counties of the present Anhui Province, and Wuyuan County of the present Jiangxi Province, were all subordinate to Huizhou Prefecture. Together they formed the then famous "one prefecture and six counties".

Huizhou Prefecture has unique geographical conditions, where mountains and hills cover nearly 90 percent of the area. Huizhou Prefecture, with the Yuling and Dazhang Mountains to its east, Zheling Mountains to its west, Jiangtan to its south and Huangshan Mountains to its north, appears geographically precipitous and hard to get to. It is recorded that Huizhou Prefecture is "isolated by mountains and the people there strongly hold to their own customs and habits."

Huizhou Prefecture, located among mountains and valleys, sets on the border between the ancient states of Wu and Chu. In the Warring States period, Wu and Chu fought each other for domination, with Huizhou area in between. Now a boundary marker read "Dividing line of Wu and Chu" still stands on the Zheling Mountains of the Zheyuan Township of Wuyuan County. Historically speaking, Huizhou Prefecture had avoided many times being involved in wars by locating seclusively and thus became a comparatively independent and safe region. On the other hand, Huizhou has been quite open. People from Huizhou traveled all around China and among them emerged the prominent Huizhou businessmen and numerous government officials. Therefore, memorial archways endowed by emperors can be seen everywhere in Huizhou, and wealthy and luxurious houses and buildings can be seen even in remote villages here.

Such unique geographical conditions helped forming the unique history of Huizhou. Nobody who has been to Huizhou can fail to feel the wealth and brilliance it once had and to learn its profound historical and cultural heritage from the steles, bridges, houses, trees, etc. Huizhou is the home town of Neo-Confucianism as well as the Peking Opera. Leading exponents in many fields of studies came from Huizhou, such as Zhu Xi, master of Neo-Confucianism who embodied the solid foundation of and the contributions from Huizhou Confucianism, and Dai Zhen, master of anti-Neo-Confucianism who symbolized the dynamic philosophical thinking of Huizhou. In Huizhou also emerged the master of abacus Cheng Da-wei of the Ming Dynasty, and famous painters in the traditional Chinese style like Jian Jiang, Wang Cai-bai and Huang Binhong, etc. There were also numerous works by scholars. Statistics show that classical works, historical works, philosophical works, and belles-lettres (the four traditional categories of a Chinese library) totaled up to 2486 volumes in the Ming and Qing Dynasties.

IV

I am a Northerner, so it's natural for me to be familiar with the local-style dwelling houses in the northern part of China. Large courtyards in the Northeast district, siheyuan (quadrangles) of Beijing and courtyards with sections of houses in Shanxi Province are nearly all bungalows of fired bricks or unfired bricks or both. But buildings in Huizhou are mainly made of stones and timbers, which directly influences, with its geographical conditions, the style and features of Huizhou constructions. According to the district annals, "the construction in Huizhou used big tree trunks which were not only cheap but also sturdy and durable. Therefore, buildings constructed in the Yuan Dynasty and other periods can be frequently seen here now, and buildings constructed in the Ming Dynasty are not rare at all. However, due to the mountains in Huizhou and its lack of fields, Huizhou buildings are mostly multi-storeyed, with high walls."

Many specialists have made precise descriptions and artistic generalizations about the style and features of the Huizhou constructions: quiet and elegant tone, unique model of walls, compact skywells and courtyards, various structures of beams, exquisite and beautiful carvings, and tasteful interior furnishings of primitive simplicity. (See Study of Huizhou Businessmen, p. 519)

"The local-style dwelling houses in Huizhou share the following things in common: well-lighted skywells and courtyards, high and steep and well arranged horse-head walls, the projecting and overlapping dougongs, and lively and vivid three carvings (wood carving, brick carving and stone carving). (See Art of Huizhou Architecture, p. 25)

Indeed, I was deeply impressed by the horse-head walls with rhythms and rhymes and the small space of the narrow skywells behind the high walls. However, the first impression was the most unique and the keenest. The door to the courtyard was big and wide, but when stepping into the doorway, I found myself standing in a cramped and dark place, which was damp and even had a stale smell. Standing in that narrow skywell and looking up, I found the sky a small rectangle and me, a well frog. It was cloudy and raining then, so the skywell didn't seem to have enough light. I thought, however, even if it were sunny, that skywell still wouldn't be as lightful as the courtyard in the North. The bedrooms were rather small, too. Rooms upstairs where women lived were even darker. Those young married women whose husbands had gone on a journey stayed in their boudoirs all day. Year in year out, leaning against the window frame and looking out, they could see nothing more than the gray, high walls and the narrow sky above. Being personally on the scene, I felt that the story told in the movie Women of Huizhou was not groundless.

Only when I trusted myself to calm observations rather than personal feelings did I get rid of the sense of being cramped and begin to savor the unique beauty of the Huizhou architecture. As cultural relics and objects for appreciation, Huizhou buildings have important historical and aesthetic value. With the passage of time, their durability and cultural glamour may all the more reveal themselves.

The horse-head wall is the most characteristic component of Huizhou folk dwelling houses. On both sides of each residence are white gables higher than the roof. The tops of the gables are arranged in different levels in accordance with the sloped roof. Both ends of the top of the fireproof walls were built in vivid shapes -- black horse-heads if the owner of the house was a military officer, and black seals if the owner was a civil official. At a distance, the dynamic horse-head walls of Huizhou hamlets resemble thousands of galloping horses.

Horse-head walls adopt grayish white colour as the key note and black as supplementary. This keynote colour of the horse-head walls is also the keynote colour of the Huizhou folk dwelling houses -- the walls are grayish white and the tiles are black. At a distance, the grayish white villages seem particularly peaceful and serene with green trees and blue waters setting them off. The color aesthetics of Huizhou folk dwelling houses is unique. The simple and plain colors create the most thought-provoking appreciation value, just as a scholar once said, "Due to several hundred years' exposition to the sun, wind and rain, the white paint on the walls has long partly peeled off and the walls show a multi-layer combination of cold and warm tones. Though the brightness and simplicity of the white colour are gone, a profound sense of history has evolved." (Yu Zhihuai: Families in the Land of Peach Blossoms, p. 27, Huangshan Publishing House)

The parvis, located between the entrance door and the principal hall, is another characteristic component of Huizhou folk dwelling houses. The parvis is not a courtyard in a strict sense. Enclosed by high walls, it is a space with a patch of sky above which gets the place more light and ventilation. It is said that the high walls were originally built for precautions against burglars, but they also caused inconveniences concerning lighting and ventilation, and made dwellers feel mentally oppressed. In order to solve this problem, a parvis was designed. A Japanese architect gave his comments on the parviss of Huizhou from an original point of view:

"The 'courtyard', frequently included in China's housing construction, is enclosed by two-storied buildings in Huizhou and thus becomes part of the house -- a 'light room'. Parvis gets its name for it is said to be the well of heaven. The central room downstairs, which opens to the entrance door and the parvis, is called 'principal hall' or 'ancestral hall'. What's going on in other rooms can not be known for they are hidden from views by the partition boards. This cross-shaped space at the bottom of the 'well of heaven' enclosed by high walls is filled with cold air emitted from stones and bricks, with breezes carrying incense smoke, with swaying chilly beams of light penetrating the wood carvings on the windows, and with the silence from complete seclusion. It is a space which seems completely secluded from the great changes taken place in China and sunk in sleep since a long time ago; it is also a space with a sense of profoundness which solidly seals up a certain Chinese culture. In this 'well of heaven' of Huizhou, there are kind of clear and serene beams of light, which I've never seen in Japanese or west European residences. Each Huizhou residence we have visited has a quiet 'parvis', but with a different shape and function. I was deeply fascinated with the marvelous 'parviss'. From them, I could see not only the glamour of Huizhou folk dwelling houses, but also the characteristics of culture in the southern area of the Yangtze River." (See Architecture of Residence, Volume 3, Institution of Japanese Architecture Trend, 1986)

With a point of view of comparative culture of architecture, the philosophical and aesthetic implication of the "parvis" in Huizhou has been caught by the above-mentioned Japanese scholar in a subtle and profound way.

The dougong in Huizhou's local-style dwelling houses is superb and incompatible. Dougong (sets of brackets on top of the columns supporting the beams within and the roof eaves without) means dou (bearing block) and gong (bracket arm, set in a bearing block; it supports a smaller block at each upraised end and often in the center). Dou is square while gong is arch-shaped. Huizhou's dougong, absorbing the quintessence of the art of dougong in the Ming and Qing dynasties, and bringing forth new ideas, has formed its unique features. It is noted that "the beauty of Huizhou dougong lies in a combination of floweriness and simplicity", and "its simplicity refers to its simple components, forthright composition and distinct lines. Huizhou dougong is dominated by simplicity while combining simplicity with complexity. Colorful paintings add luster and glamour to its simple structure." (See Art of Huizhou Architecture, pp. 102-106) While appreciating dougong of various shapes under the eaves, I cannot help wondering if it is necessary to build dougongs so exquisitely, especially at this unnoticeable place. However, I also cannot help admiring Huizhou people's minute and subtle sense of beauty, displayed in every way and everywhere possible. They turn dougong, a construction component, into a piece of art even under the eaves.

V

Yet another characteristic of Huizhou-school architecture is the Three Carvings, that is, woodcarving, stone carving and brick carving.

Huizhou-style construction and Huizhou carvings exist in harmony and are inseparable. Where there is Huizhou-style construction, there must be Huizhou carvings. The exquisite Huizhou carvings can be seen everywhere -- folk residences, ancestral halls, temples, memorial archways, towers and pagodas, etc. -- and on each component of Huizhou construction -- beams, columns, architraves, dougongs, doors and walls. Huizhou carvings emphasize the aesthetic sense of the construction components while maintaining their practicality.

There are three kinds of Huizhou carvings.

First, woodcarvings. After you step into the doorway, your eyes will be filled with fine woodcarvings on beams, columns, dougongs, quetis, window frames, furniture, screens and utensils in the study. And thus various categories of carvings came into being, such as carvings on quetis, dougongs, architraves, hook decorations, roofs, eaves, frieze panels and crescent beams. Together they create a dense cultural atmosphere and elegant environment in a house. The fine carvings can be seen everywhere and thus give pleasure to the dwellers' eyes and mould their temperament.

I had seen many woodcarvings before, but none could have impressed me more deeply than the woodcarvings in Huizhou. Groups of woodcarving images, and series woodcarving stories from the Three Kingdoms, etc., are all too fabulous (See picture 1). The engraving techniques used in Huizhou carvings are skillful and exquisite. Take the carving Emperor Su-zong Entertaining the Officials of the Tang Dynasty, in Chengzhi Hall in Xidi Village, as an example.

The carving, one chi tall and six chi wide, consists of five or six tables arranged in a single-line formation, at which there are about 30 people taking different gestures and with different expressions. Some are playing the qin (a seven-stringed plucked musical instrument in some ways similar to the zither); some are playing weiqi (a chess game played with black and white pieces on a board of 361 crosses); some are drawing and others are writing calligraphy; there is even one person picking his ears. This piece of carving is not only interesting, but also complete in its composition. Its carving technique has reached the acme of perfection by creating seven or eight layers on a wooden board about five or six centimeters' thick.

The photos in this album show woodcarvings representing the peak of perfection, like "the queti of nine lions" (See picture 2) of the Yuqing Hall, and other rare masterpieces, like the woodcarvings in the Hu Clan Ancestral Hall and the Yu Clan Ancestral Hall, and the woodcarving of A Pair of Phoenixes Facing the Sun in the Zhifeng Hall.

Second, brick carvings. Brick carving is specially made of dark-blue bricks carved for the use of shielding above the gates and windows, and arches over gateways (See picture 3). Elegant brick carvings are used to adorn the exterior of a building and they are Huizhou's unique creation. Not as pliable as wood and as hard as stone, the bricks are still rough though they are smooth bricks polished with waterstones. However, people in Huizhou have turned them into art pieces. The brick carvings on gate shielding, which I saw in Bei'an Village, were stunning, with a precise composition, protruding patterns, views of gardens and groups of figures carved in the center, which are extremely exquisite and sometimes having even nine layers -- more layers than wood carvings. The solemn and elegant brick gate-towers like those of drooping flowers, of scribed boards style and archways, are even more tasteful.

The third is stone carving. People in Huizhou often used stone carvings in their residences, such as those on the lion-pattern drum-shaped bearing stones, entrance walls, fences, steps, pedestals and tablets, and in places like temples, ancestral halls, steles, memorial archways, bridges and pagodas. In the Erjia Ancestral Hall of Chaji Village, I saw exquisite pedestals and drum-shaped stones.

In Yixian County, stone carvings can be seen everywhere. The thing fascinated me most was that those stone carvings often represented their owners' spiritual sustenance. It is said that Hu Wenzhao, owner of the West Garden in Xidi Village, was once an ambitious official, and his having the story of Jiang Ziya Fishing carved on the lintel of the gate might have indicated his wishes of resigning and living in seclusion.

Fretted windows were originally designed for better ventilation and lighting, but people in Huizhou turned them also into works of art. Those windows with stone-carved images like "pines and stones", "bamboo and plum tree", "tree leaves", "lute, chess, calligraphy and painting" or the pattern of "Three Mates in Bitterly Cold Days—pine tree, bamboo and plum tree" show the owners' leisurely and carefree mood. The image of leaves may indicate that a person residing elsewhere would finally return to the land of his ancestors (See picture 4). The pattern of "Magpie on the Plum Tree" expresses hopes for a happy life, and that of "Carp Leaping over the Dragon Gate" refers to the yearning for a brighter future (See picture 5). All these images and patterns have shown one aspect of the cultural taste of the Huizhou architecture.

The inscription on steles is another category of stone carvings. It might have been neglected, but it does have important historical and artistic values. You can see steles with inscriptions everywhere -- in the fields, in front of the villages, by the riverside, in the courtyards, beside the houses and at the riverbank beside a bridge. There are many kinds of inscriptions -- historical records, biographies, and records of the process of constructions, which have become precious historical records as well as works of art, collections of calligraphy and the treasury of poetic and literary works. For example, the article Missing Wang Xinlan in Spring When Passing Leigang preserved in the South Lake Academy in Hongcun Village was inscribed on a stele in the Ming Dynasty. And the two stone carvings of "Pine Tree, Plum Tree and Bamboo", preserved in the Bureau of the Preservation of Cultural Relics of Yixian County, are precious works of art presenting the harmony between the poem and the drawing.

People may ask why Huizhou people put so much emphasis and time and energy on the interior ornament, from the design of doors and windows to the carvings on beams and columns, in such a meticulous way. It is recorded that there were limitations and regulations on the construction of houses for ordinary people as early as in the Ming Dynasty. For example, exceeding the limitations on the scale of construction was not allowed, and the use of resplendent and magnificent ornaments was also banned. Those who would violate these regulations might even be executed. Therefore, Huizhou businessmen, within the limits permitted by laws, tried all possible means ornamenting the interior of their yards and rooms. They concentrated on the layout of their houses, made the houses more compact and solid, and used exquisite ornaments to make them more beautiful and tasteful. Thus, the houses could become the places where they could spend their remaining years in happiness after resignation, as well as the objects with which they could compete with their fellow villagers in wealth. Thus, the labor-intensive woodcarvings, brick carvings and stone carvings came into being, and became the unique ornamental style of Huizhou construction.

VI

The glamour of Huizhou-school architecture comes not only from the style and features of the construction itself, but also from the hamlets formed by groups of constructions and the harmony between the hamlets and the nature.

In order to better understand the overall conception and planning of these hamlets, let's look at the characteristic of "people of a clan living together" in Huizhou hamlets.

"People of a clan living together" was the most prominent feature of ancient Huizhou hamlets. Huizhou Prefecture was one of the regions where family system prevailed. Since the Han Dynasty, those hamlets based on ties of blood and formed by people from the same clan came into being, and the people there multiplied. According to historical records, the custom of "people of a clan living together" came to exist since people from central China began to migrate to Huizhou. When numerous people from the North migrated to the South by clans and arrived at Huizhou region, they usually settled by mountain passes or in strategically located places which were difficult of access but easy to guard. Then they formed hamlets by clans and multiplied. For example, Xidi Village of Yixian County was the settlements of the clan of Hu. According to the records in Hu Clan Genealogy, the surname of Hu's ancestor was Li and he was descendant of Emperor Zhao Zong of the Tang Dynasty. Hu Shiliang, descendent of the fifth generation of the clan, discovered Xidi. Since then, the Hus have been living in Xidi for nearly 300 years and multiplied for dozens of generations. "People of a clan living together" made the overall planning of a hamlet possible.

By the middle of the Ming Dynasty, hamlets in Huizhou had formed their unique patterns and landscapes. After the mid-Qing Dynasty, hamlets in Huizhou reached their heyday. The construction pattern of the folk residences achieved maturity and perfection, and the overall planning of the hamlets was harmonious and reasonable. Other constructions like memorial archways, ancestral halls, bridges, gardens, ponds and academies had all been included in the overall environmental plan of the hamlets.

Several hundred years later, some hamlets were still preserved in good condition in Huizhou, like Xidi, Hongcun, Nanping, Guanlu of Yixian County, Chengkan, Xucun, Tangyue of Shexian County, Wangkou of Wuyuan County and Yingzhou of Jixi County. These hamlets have become exhibiting centers of Huizhou-style construction, tourist attractions of beautiful landscape, and museums of historical relics.

In order to achieve harmony among dwellers, people from the same clan lived together; in order to achieve harmony between Heaven and man, the location of settlements must be carefully chosen. Therefore, considering the planning of the hamlets and taking aesthetics into account, people of Huizhou always chose valleys, ridges, mountain passes, water-outlets, brooks, ferry crossings and riverside when they built their hamlets.

As has been summarized in Art of Huizhou-school Architecture, "Huizhou hamlets are located either at the foot of mountains -- at mountain glens or passes, or by the riverside -- at the bends of rivers, ferry crossings or the branches of rivers. Hamlets and buildings scatter all over Huizhou. They are in different shapes and also in harmony and order. Their patterns are regular. Overlooked from high places, the shapes of Huizhou hamlets can be summarized as follows -- an ox horn shape like folk dwelling houses in Xikeng, Wuyuan County; a bow shape like Taibaisi residences in Wuyuan County; a ribbon shape like Gaosha folk residences in Wuyuan County; a wave shape like folk residences in Xidi, Yixian County; an ox stomach shape like local-style dwelling houses in Hongcun, Yixian County; a nebulous cluster shape like Qiankou local-style dwelling houses in Shexian County; a winding shape like Jiangcun local-style dwelling houses in Shexian County; and a semicircle shape, shape of a square seal, an arc, a straight line, and dots, etc."

Hongcun in Yixian County is a typical case. Hongcun began to exist in the Southern Song Dynasty. At the very beginning, there were only 13 houses in the village, which was the settlement of Wang's family. It is said that Wang's family did not prosper for several generations, so they thought the reason might be that they hadn't made good use of the area's fengshui--geomantic omen. Therefore they sent for the famous geomancer, He Keda of Xiuning County, to examine the terrain. He traveled all over Hongcun and made a thorough investigation of the way the mountains extended and the rivers flew. He believed that the terrain of Hongcun resembled a crouching ox, so he decided to make the overall planning for Hongcun according to the functioning system of an ox. It was very rare at that time in China to make an overall planning for hamlets, which was a unique feature of Huizhou-style construction. In the course of over a hundred years, Hongcun, by using a natural spring, constructed a pond in the shape of half-moon as "the stomach of the ox", dug a 400-metre-long channel as "the intestines of the ox", and built four bridges over Yushan brook in the west of the village as "the four hoofs of the ox". Thus a hamlet in the shape of an ox came into being, with mountains as the ox's head, trees as its horns, houses as its body, and bridges as its hoofs. Later, a small artificial lake, South Lake, was constructed in the south of the village, as another "stomach of the ox". Hongcun, from then on, became a beautiful landscape as well as a serene and poetic place. I have visited many hamlets, in the south, or in the north; however, Hongcun, a fairyland of peace and beauty, is the only place where I couldn't tear myself away. I was overwhelmed with admiration for our ancestors' poetic sense of beauty and wisdom, and their philosophy about the harmony between Heaven and man.

The custom that people from the same clan lived together made ancestral halls in Huizhou extremely spectacular. Ancestral halls can be seen in nearly every village. Patriarchal clan -- group based on blood ties -- has been the cornerstone of China's social organizations and the concept of patriarchal clan system throughout the long years of Chinese history. The worship of ancestors and belief in patriarchal clans brought about the construction of ancestral halls as well as the spirits of and feelings about ancestral halls. It deeply penetrated the pattern and style of Huizhou-style construction. It is a construction system on the surface and a mental power deep inside.

Huizhou-style buildings like folk residences, memorial archways, pagodas, pavilions, academies and gardens were all designed in a pattern with ancestral halls in the center.

Ancestral halls naturally became the places where people worshiped their ancestors, for the halls embodied the worship of ancestors. Take Baolun Pavilion as an example. There collected imperial mandates and edicts to the Hu's clan in Chengkan Village, which obviously showed the ancestral halls' concept of the worship of ancestors. At the same time, the memorial archways in front of the ancestral halls or hamlets were also aimed at advocating the clan's celebrated status and brilliant achievements, while pagodas and pavilions were also symbols of the clan's dignity, beliefs and benevolence. Ancestral halls formed the system of Huizhou-style construction. However, the spirits the ancestral halls had embodied were patriarchal and feudal.

Take Xidi Village as an example. It is recorded that at its heyday, the Hu's clan owned 34 ancestral halls, including general ancestral halls, head ancestral halls, branch ancestral halls, and family ancestral halls. It is said that Xidi's ancestral halls stood like a forest and Xidi, a world of ancestral halls. Both the brilliance and costs of those ancestral halls could make one stunned. According to the records from a stele in one of the ancestral halls, 6940 liang of silver, contributed by all the branch halls and the rich families, was spent on the construction of that ancestral hall. The penetration of the patriarchal concept could be viewed from this.

VII

It sometimes makes me wonder why there are so many Huizhou-style buildings and nice hamlets in an environment enclosed by mountains. There may be many reasons but one thing is for sure: the emergence of Huizhou merchants was the economic foundation of the prosperity of Huizhou's architectural culture.

Huizhou merchants refer to the groups of businessmen from Huizhou, based on clan relationship. They rose in the middle of the Ming Dynasty, and declined in the last years of the Qing Dynasty. It is said "No Huizhou merchants, no towns and cities". They had a decisive position in the economical development in the Ming and Qing periods. They mainly went in for and almost controlled the salt and financial businesses in China at that time. It was easy for them to become rich and no other businessmen could compete with them because they were protected by the government and exempted from taxes. They raked in exorbitant profits but they didn't want to invest on the production of commercial goods. With the money they were busy putting up buildings in their home towns.

Huizhou merchants usually operated away from home. When they returned to build their homeland, they brought with them various styles of construction of different places. For example, the gardens in Suzhou and Yangzhou were then very famous, so Huizhou merchants brought back home the pleasure of building gardens of Suzhou and Yangzhou style, and hence forming the custom of building gardens in Huizhou, too.

VIII

Huizhou merchants' wealth was no doubt one of the reasons for the brilliance of Huizhou architectural culture, but the profound connotation of this architectural culture also demonstrated the far-reaching and extensive sources of Huizhou culture.

Some scholar has pointed out that ancient Huizhou hamlets stressed the importance of humanism, which was showed as follows: first, the beautiful landscape molded a person's temperament and encouraged people to make progress; second, those hamlets educated people with culture and civilization; third, the hamlets gathered people by ceremony and propriety. Indeed, Huizhou architecture contains the inside details of the Confucian culture and embodies a scholarly and refined bearing. In Chapter 22 of The Scholars by Wu Jingzi, there is a paragraph of description of a Huizhou merchant's residence in Yangzhou, which resembled the pattern of Jing'ai Hall in Xidi Village; even the couplets of the hall were the same. It is said that businessmen as Huizhou people were, they tried to imitate and follow scholars' manners. Merchants worked for profits and scholars sought for fame and reputation. When people failed to become scholars, they turned to being engaged in business. When they became successful businessmen, they would rather be in favour of scholars for the sake of their descendants.

Of course, those descriptions touched only the surface of things, but through the surface of things, we can understand the profound connotation and tradition of Huizhou culture. Huizhou was not only a land of businessmen, but also a land of literary people. Since the Song and Yuan Dynasties, Huizhou demonstrated its developed education and large numbers of talented people. The reason why a remote mountainous area like Huizhou could become a land of talented people was that, during many war periods, rich people, talented people and people with high social status poured in Huizhou to avoid the chaos of wars. It is said that even in a village with only ten households, there would be some literary people. What's more, Huizhou merchants had the habit of doing academic studies while doing business. They either did business first or did academic study first, or did both at the same time. During the Ming and Qing Dynasties, giving lectures had become a common practice in the six counties of Huizhou Prefecture. Just then, there were 54 academies, the most famous among which was Ziyang Academy. Those academies nurtured a great number of talents for Huizhou. According to statistics, they had turned out 298 Jurens and 392 Jinshis in the Ming Dynasty, and 698 Jurens and 226 Jinshis in the Qing Dynasty. (Juren, a successful candidate in the imperial examinations at the provincial level in the Ming and Qing dynasties; Jinshi, a successful candidate in the highest imperial examinations.) On the one hand, there was the scholarizing tendency of the Huizhou merchants; on the other hand, the officialization of the Huizhou merchants. Statistics show that in the Qing Dynasty, Shexian County alone had nurtured four Grand Secretaries of the Cabinet, seven ministers, twenty-one vice directors of one of the Six Boards, seven censors of Censorate, and fifteen secretaries of the Grand Secretariat. Dr. Hu Shi, a famous Chinese scholar, once said, "we Huizhou people could be the most advanced in culture and education in every dynasty ... since the middle ancient times, the fact that some scholars could achieve relatively high status in academic circles of the middle ancient times was not accidental."

However, we should also pay attention to the negative impact of the feudal ethical code upon Huizhou architecture. "Beauty chairs" was one example. According to the feudal ethical code, girls had to stay upstairs and could not go downstairs as they pleased. Therefore, in some houses, a row of chairs were arranged in the living room upstairs, by the side of the parvis, so that the girls could observe what was going on downstairs and outside the courtyards. Thus, the row of chairs were given the name "beauty chairs".

It was with this cultural tradition that Huizhou-style constructions prospered and formed their unique features.

IX

In the end, let's talk about Mr. Fan Yanbing, the chief editor of this book.
In the opinion of people outside Mr. Fan's circle, it must have taken a department

of the government, a scientific research institution with capabilities, or a publishing organization with vision, to organize and publish such a monumental work; and it must have taken a specialist in Anhui local-style dwelling houses or a prestigious and well-known scholar to edit this book. I'm afraid that these assumptions are quite wrong. Mr. Fan Yanbing completed this book, all depending on his deep love of Huizhou local-style dwelling houses and photography, on his accumulation and energy, on his own investments, and on the investigation and photographing carried out by him and his friends. Therefore, the success of this book is a miracle taken place during China's reform and opening-up, and it is a pioneering work deserving recognition.

I have learned of all kinds of hardships Yanbing had encountered during the ten years of his editing this book, from Yanbing himself and his friends. He said, "this book is the fruit of my painstaking labor and the fruit of the passions of many people whom I know or do not know."

Mr. Fan Yanbing graduated from Anhui Institute of Architectural Industry in 1984, and naturally he was engaged in architecture. However, it was by chance that he became interested in Huizhou local-style dwelling houses. After he was assigned to work in Anhui Provincial Research and Design Institute for Urban and Rural Planning, he had the opportunity to take part in the Institute's investigation into Anhui local-style dwelling houses. As a new comer in the Institute, he was quite unfamiliar with Anhui folk residences at the beginning. However, during the next twenty days' investigation, he visited nearly all six counties of the ancient Huizhou Prefecture, except Wuyuan County in Jiangxi Province. He was deeply impressed with those Huizhou-style constructions, and was amazed by the unique aesthetic features of Huizhou local-style dwelling houses, their profound cultural connotation, and their being so intactly preserved. Such brilliant cultural relics fascinated him and thus flew into his heart and his life.

As he once said, the more he was engaged in Huizhou local-style dwelling houses, the more he felt mentally attached to them. However, it was after he had read the books, Introduction to Chinese Residences by Liu Dunzhen and Huizhou Dwelling Houses in the Ming Dynasty by Zhang Zhongyi et al, that he wished he would photograph all those houses, for he found that those Huizhou local-style dwelling houses in the Ming Dynasty mentioned in the books were no longer there when he tried to visit them, which deeply hurt him. Staring blankly at the ruins and remains, and witnessing those houses being torn down and the remains of doors, windows, bricks and beams being sold as cultural relics, he felt extremely painful. He promised in 1992 that he would photograph all those valuable buildings which were still standing on the land of south Anhui Province, and make complete archives of the historical images of Huizhou folk residences.

Yanbing has been working in Guangdong Province since 1992, but he concerns about Huizhou all the time and visits Huizhou several times every year. He was keenly aware that to carry out such a magnificent plan, his own efforts wouldn't be enough.

Therefore, he made friends with and learned from those working in the local Bureau of Culture, and those local specialists. Every time he returned to Huizhou, he would visit those friends and seek their advice. In this way, he has opportunities to visit not only those folk residences open to exhibition, but also those rarely known to the outside world.

Yanbing particularly introduced one of his friends in Huizhou -- Mr. Xu Zhiyong. Mr. Xu is quite familiar with Huizhou local-style dwelling houses. He understands Huizhou dialects and can drive. Yanbing said, "Mr. Xu Zhiyong has accompanied me for nearly ten years investigating Huizhou local-style dwelling houses, and has rendered a great service." What's more, several people were also moved by Yanbing's morale and joined in this plan. They are -- the other two photographers of this book -- Mr. Zhang Zhenguang, senior architecture photographer from China Architecture & Building Press, and Mr. Wu Guangmin, a young photographer from Anhui Province. They have both done their utmost and rendered a great help for the publication of this book.

Yanbing's morale has also moved the local people. Moved by Yanbing's magnificent plan, a villager in Hongcun found him and supplied him with information of some buildings rarely known. Bao'ai Hall was recorded to have a large scale of construction with 108 rooms. However, Yanbing and his friends found it in ruins when they went to photograph it. When the owner of the hall, an old man, learned of Yanbing's program, he drew a plan of the 108 rooms according to his memories. Cases like this are too numerous to mention.

Yanbing, with his passion for the cultural relics of his motherland, and out of a strong sense of responsibility, completed not only such a valuable work for his motherland and its people, but also the molding of his own cultural character. In the great tide of commodity economy, and in the kind of social atmosphere in which studies of arts are ignored, Yanbing's devotion of his financial and material resources, and manpower as well, to the sacred cause of the accumulation of Chinese culture, is extremely precious. I myself have also been deeply moved.

I would like to conclude the article by extending my best wishes to Mr. Fan Yanbing, chief editor of the book. I hope he will continue his cause and I also hope readers will enjoy the book.

First draft: November 2001
Final version: January 2002
In Beijing

目 录
CONTENTS

序言（罗哲文）
Preface (Luo ZheWen)

徽派建筑漫记（田本相）
Some Notes about Architecture of Huizhou School (Tian BenXiang)

古祠堂
Ancient Ancestral Halls

宝纶阁 Baolun Pavilion	2
北岸吴氏宗祠 Ancestral Hall of the Wu Clan in Bei'an	8
周氏宗祠 Ancestral Hall of the Zhou Clan	14
棠樾鲍氏支祠 Branch Ancestral Hall of the Bao Clan in Tangyue	18
棠樾清懿堂 Qingyi Ancestral Hall in Tangyue	24
郑村郑氏宗祠 Ancestral Hall of the Zheng Clan in Zheng Village	28
司谏第 Sijiandi-Residence of the Remonstrator	34
三槐堂 Sanhuai Hall-Hall of Three Chinese Scholartrees	36
进士第 Jinshidi-Residence of Jinshi	40
经义堂 Jingyi Hall	44
俞氏宗祠 Ancestral Hall of the Yu Clan	50
成义堂 Chengyi Hall	58
龙川胡氏宗祠 Ancestral Hall of the Hu Clan in Longchuan	64
贞一堂 Zhenyi Hall	76
馀庆堂 Yuqing Hall	80
叶奎光堂 Ye Kuiguang Hall	86
敬爱堂 Jing'ai Hall-Hall of Respect	92

古牌坊
Ancient Gateways

许国石坊 Xu Guo Memorial Stone Gateway	102
同胞翰林坊 The Brother Hanlin (Members of the Imperial Academy) Gateway	106
棠樾牌坊群 Group of Gateways in Tangyue	108
四世一品坊 Sishiyipin (Highest Court Officials for Four Generations) Gateway	110
四面四柱石坊 Four-side and Four-column Stone Gateway	112
昌溪木牌坊 Changxi Wooden Gateway	114
奕世尚书坊 Yishi Shangshu (Ministers) Gateway	116
胡文光刺史坊 Hu Wenguang Cishi (Provincial Governor) Gateway	118
方氏宗祠坊 Gateway of Fang Clan Ancestral Hall	120

古 塔
Ancient Pagodas

长庆塔 Changqing Pagoda	124
新州石塔 Xinzhou Stone Pagoda	125
巽峰塔 Xunfeng Pagoda	126
丁峰塔 Dingfeng Pagoda	127
富琅塔 Fulang Pagoda	128
古城塔 Gucheng Pagoda	129
神皋塔 Shengao Pagoda	130
下尖塔 Xiajian Pagoda	131

古 桥
Ancient Bridges

洪桥 Hong Bridge	134
彩虹桥 Rainbow Bridge	138

古 亭
Ancient Pavilions

文昌阁	Wenchang Pavilion—The Pavilion of Flourishing Culture	144
绿绕亭	Lürao Pavilion—The Pavilion of Greenery Winding	146
沙堤亭	Shadi Pavilion—The Pavilion of Sand-Bank	148
善化亭	Shanhua Pavilion—The Pavilion of Mercy	150
魁星阁	Kuixing Pavilion—The Pavilion of God of Literature	152

古书院
Ancient Academies

竹山书院	Zhushan Academy—The Academy of Bamboo Hill	156
古紫阳书院	The Ancient Ziyang Academy	158
文庙书院	Wenmiao Academy—The Academy at the Temple of Confucius	160
考棚	The Examination Chambers	164
南湖书院	Nanhu Academy—The Academy by the South Lake	166

古民居
Ancient Folk Residence

许国相府	Residence of Premier Xu Guo	172
汪宅	Residence of the Wangs	174
许氏大院	The Compound of the Xus	178
黄宾虹故居	Former Residence of Huang Binhong	180
老屋阁	Old House	184
巴慰祖故居	Former Residence of Ba Weizu	188
吴建华宅	Wu Jianhua Residence	192
方观田宅	Fang Guantian Residence	194
方文泰宅	Fang Wentai Residence	196
德庆堂	Deqing Hall	200
遵训堂	Zunxun Hall—The Hall of Following Teachings	204
保艾堂	Bao'ai Hall	208

汪氏家戏台 The Stage of Wang Clan	210
程大位故居 Former Residence of Cheng Dawei	214
程氏三宅 Three Houses of the Cheng Clan	218
金馀庆堂 Jin Yuqing Hall	226
聪听堂 Congting Hall—Hall of Sharp Hearing	230
云溪别墅 Villa of Yunxi	236
俞氏客馆 Guest House of the Yu Clan	240
胡寿基宅 Hu Shouji Residence	244
胡适故居 Former Residence of Dr. Hu Shi	252
胡开文纪念馆 Memorial Museum of Hu Kaiwen	256
石磐安宅 Shi Pan'an Residence	258
跑马楼 Paoma Tower—Horse Racing Tower	262
瑞玉庭 Ruiyu Hall—Lucky Jade Hall	266
西园 West Garden	270
东园 East Garden	276
履福堂 Lüfu Hall—Hall of Experiencing Happiness	280
膺福堂 Yingfu Hall—Hall of Receiving Blessing	282
笃敬堂 Dujing Hall—Hall of Deep Respect	286
大夫第 Dafu Di—Senior Official Residence	290
笃谊堂 Duyi Hall—Hall of Deep Friendship	296
承志堂 Chengzhi Hall—Hall of Behest-Inheriting	300
碧园 Bi Yuan—The Green Garden	314
八大家 Residence of Eight Prominent Families	316
迎祥庭 Yingxiang Hall	328
冰凌阁 Bingling (Icicle) Hall	332
木雕楼 Wood-carving Building	336

再版后记
Postscript for the Revised Edition — 342

附：主要参考书目 — 344

中国徽派建筑
HUIZHOU SCHOOL ARCHITECTURE OF CHINA

古祠堂

Ancient
Ancestral
Halls

宝纶阁

歙县呈坎村

宝纶阁位于徽州区呈坎乡呈坎村北端，原名"贞静罗东舒先生祠"，坐西朝东，背山面水，占地3300多平方米，气势宏大，据《贞静罗东舒先生祠堂记》记载，该阁始建于明嘉靖年间，先建寝殿（1540年），历经七十余载，毁坏严重，明万历四十年（1612年）又重新扩建，至万历四十五年（1617年）落成。分别是嘉靖和万历两度建造，故形式和风格迥异。该阁用于珍藏历代皇帝赐予呈坎罗氏的诰命、诏书等恩旨纶音及家族议事。阁为二进歇山顶，祠堂前沿溪照壁面宽29米，呈八字形，进而是棂星门，大门绘有彩绘门神，其后左右建有两碑亭，立《祖东舒公祠堂记》碑碣于其中。再进是仪门，上悬罗哲文先生所题"贞静罗东舒先生祠"匾额，穿过仪门即为宽大的天井，天井当中是甬道，两旁各有庑廊，两庑廊阶前临天井池处均有雕刻精美的石雕栏板。进入第二进大厅，大厅名"善厅"，享堂悬有巨大匾额，上书"彝伦攸叙"四个大字，为明万历进士董其昌手书。寝殿高出前堂1.6米多，殿前是一道浮雕石刻栏板。栏板雕刻精美，每块图案各异。寝殿是供奉祖先神位的所在，也是整个祠堂的精华所在，共十一间，十根檐柱采用琢成讹角的方形石柱,檐下正中悬着吴士鸿手书的匾额"宝纶阁"。

宝纶阁全景 A full view of Baolun Pavilion

宝纶阁剖面图 Section view of Baolun Pavilion

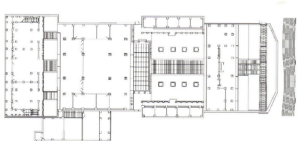
宝纶阁底层平面图 Plan of the first floor

Baolun Pavilion

the town of Chengkan, Shexian County

Baolun Pavilion, formerly named The Memorial Temple of Mr. Luo Dongshu of Zhenjing, is located in the north of Chengkan Village in the town of Chengkan, Huizhou District. Facing east towards the water, it covers an area of over 3300 square meters, lying against the mountains in an imposing manner. According to Annals of the Memorial Temple of Mr. Luo Dongshu of Zhenjing, the Qindian or Ceremonial Hall was first constructed in the Jiajing reign of the Ming Dynasty (1540). The pavilion was seriously damaged in the following period of about seventy years, and was rebuilt and extended in the fortieth year in the Wanli reign of the Ming Dynasty. The reconstruction was finished in the forty-fifth year of the Wanli reign (1617). Since its two separate constructions in the Jiajing & Wanli reigns, the two parts are diametrically opposed in shape and style. In the pavilion were stored the imperial mandates and edicts conferred by the emperors and the records of the clan discussions. The pavilion has the xieshan, or Chinese hip-and-gable roof, with two sections of houses. Along the brook, a 八-shaved screen wall, 29 meters in width, stands in front of the temple. The Lingxing Gate is decorated with colored door-gods. On both sides behind the gate stand two stele pavilions. The stone tablets of inscribed Annals of the Memorial Temple of Ancestor Mr. Luo Dongshu stand there. An inscribed board of "The Memorial Temple of Mr. Luo Dongshu of Zhenjing" written by Mr. Luo Zhewen, is hung above the Yimen or Ceremonial Gate. The paved path goes across the spacious courtyard with corridors on both sides. There are delicately sculptured stone railing panels along the corridors. The second main hall is named Shan Hall. There hangs a huge board of four characters "Yi Lun You Xu", written by Mr. Dong Qichang, a Jinshi in the Wanli reign of the Ming Dynasty. With delicately carved panels, Qindian is 1.6 meters higher than the antechamber. It is the place to lay the memorial tablets of the ancestors and also the essence of the whole memorial temple. It is an eleven-bayed hall with ten stone square columns cut into the round-off corners. The inscribed board of "Baolun Pavilion" written by Wu Shihong is hung right in the center below the eaves.

宝纶阁祠堂正门 Main entrance

宝纶阁前院 Front courtyard

宝纶阁梁架彩绘 Colour paintings on the beams

吴士鸿手书"宝纶阁"匾文 Inscribed board with characters written by Wu Shihong

宝纶阁

中/宝纶阁石阶栏板及石狮 Stone frieze panels of steps and stone lions

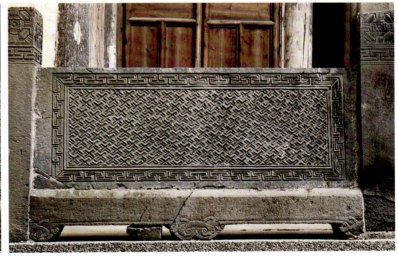

宝纶阁石栏板 Stone frieze panels

宝纶阁寝殿前天井 Parvis in front of Qindian or Ceremonial Hall

宝纶阁享堂 Xiangtang or Reception Hall

宝纶阁一进大院 Courtyard of the first section

北岸吴氏宗祠 \ 歙县北岸村

北岸吴氏宗祠位于歙县北岸村。建于清道光年间（1826年）。三进三开间。门厅为五凤楼建筑，八字墙须弥座石刻与檐下砖雕、博缝板木雕均极华美。中进享堂月梁、金柱粗硕宏大，为徽州之罕见。檐柱前有"黟县青"石栏，望柱头刻石狮，栏板上镌刻杭州西湖风景，洗练精致。寝殿台基前立石栏与两边台阶垂带石栏板相接，寝殿前栏板镌刻"百鹿图"场面，群鹿隐现于山林间，千姿百态，栩栩如生。中进后廊天井栏杆，由13块栏板组成，望柱上饰石狮，栏板上镌刻礼器，亦极精致。下为蓄水池。寝殿右侧墙上置一砖雕神龛，三间殿宇式，雕作细腻。左侧墙上竖嵌修祠碑记一方，右侧墙上同样竖嵌功德碑记一方。

北岸吴氏宗祠大门 Entrance

Ancestral Hall of the Wu Clan in Bei'an

Beian Village, Shexian County

Ancestral Hall of the Wu Clan is located in Beian Village, Shexian County. It was set up in the Daoguang reign of the Qing Dynasty (1826). It is three-bayed and three-sectioned. The gate-hall is of the style of five-phoenix structure, with magnificent stone carvings on the xumizuo base of the screen wall, gorgeous brick carvings below the eaves and beautiful woodcuts of the gable boards. The crescent beam and principal columns in the main hall of the middle section are thick and grand, which is an exception in Huizhou District. There are railings made of "Yixian Black Stone" in front of the eave columns. On the baluster capitals there are lion-shape sculptures and on the railing panels there are carvings of the scenery of the West Lake in Hangzhou, succinct and delicate. The stone railings in front of the platform of Qindian are connected with the festoon railing panels along the steps on both sides. One Hundred Deer Design is engraved on the railing panels in front of the hall. The railings in the courtyard of the middle section consist of thirteen carved panels, with stone lions on the top of the baluster capitals. Below is a storage pond. On the right wall of Qindian is a brick-carved shrine, which is of three-bayed palace-shape. On the left wall is embedded a tablet of inscription about the construction and on the right wall a tablet of merits and virtues.

北岸吴氏宗祠寝殿前天井水池　Pond of the parvis in front of Qindian

中国徽派建筑　古祠堂

北岸吴氏宗祠寝堂中的碑记
Tablet inscription in Qindian

北岸吴氏宗祠寝堂中的碑记
Tablet inscription in Qindian

北岸吴氏宗祠寝殿中砖雕神龛
Brick carved shrine in Qindian

北岸吴氏宗祠一进天井　Parvis of the first section

北岸吴氏宗祠

北岸吴氏宗祠望柱石狮　Stone lions on the balusters

中国徽派建筑　古祠堂

北岸吴氏宗祠寝殿前栏板石雕"百鹿图"局部　Part of Stone carvings of "One Hundred Deer"

北岸吴氏宗祠享堂栏板石雕"西湖风光"　Stone carvings of West Lake Scenery in Xiangtang

北岸吴氏宗祠寝堂前栏板石雕"百鹿图"　Stone carvings of "One Hundred Deer"

北岸吴氏宗祠寝殿前栏板石雕"百鹿图"局部　Part of Stone carvings of "One Hundred Deer"

北岸吴氏宗祠享堂栏板石雕"西湖风光"　Stone carvings of West Lake Scenery in Xiangtang

周氏宗祠

歙县昌溪村

周氏宗祠位于歙县昌溪村。建于清乾隆年间，坐北朝南，背山面水，占地面积1300平方米。门前有一广场，门厅为五凤楼、八字墙须弥座石刻，砖木结构，分别由门厅、享堂、祀堂和天井组成，门楼的两边有侧门，平常由侧门进出，大门两侧有"黟县青"石鼓一对，雕饰有龙凤呈祥、麒麟送子等图案。进门为前进，左右两厢，过天井即为正厅，两庑及正厅沿天井的柱子和柱础均为石柱和石础，天井规模宏大，是家族举行庆典的场所。享堂的正中悬挂着"六顺堂"的匾额，后进是祀堂，比前堂高出近一米，天井中有一长方形水池，水池四周石柱和栏板上，石柱望头为荷花头，石拦板为巨大的青石板，水池两边有台阶到祀堂，祀堂为二层，檐口悬挂着"祖德显丕"的匾额，底层是祭祀堂，系供奉昌溪周氏祖宗神位所用，两侧有楼梯可上二层。

周氏宗祠大门　Entrance

Ancestral Hall of the Zhou Clan

Changxi Village, Shexian County

Ancestral Hall of the Zhou Clan is located in Changxi Village, Shexian County. It was built in the Qianlong reign of the Qing Dynasty. Facing south towards the water, it covers an area of over 1300 square meters with a square in front of it. The entry hall is of five-phoenix style with the stone carved xumizuo bases of the splay walls. The whole hall, which is brick-and-timber construction, consists of the entry hall, reception hall, sacrificial hall and parvis. The side doors on each side of the gate hall are for daily use. On both sides of the gate are a pair of stone drums made of black stones from Yixian County, with designs of dragon and phoenix bringing prosperity and unicorn bestowing a son. The gate is followed by the first section with two wing rooms on each side and a courtyard in front. The pillars and the bases around the yard are made of stones. The main hall, broad in scale, is the place for the clan to hold celebrations. In the center of the main hall is hung an inscribed board of "Liu Shun Hall". The last section is the sacrificial hall, which is about a meter higher than the first hall. A square pond is in the middle of the yard. Stone lotuses are on the top of the baluster capitals around the pond and the railing panels are large black stone boards. There are steps on both sides of the pond leading to the two-storied sacrificial hall. An inscribed board of "Zu De Xian Pi" is hung below front eaves, which means the lofty virtues of the ancestors. On the first floor is the sacrificial hall where memorial tablets of the ancestors of the Zhous in Changxi are placed. There are side stairs leading to the second floor.

周氏宗祠正面 Facade of the Ancestral Hall

中国徽派建筑　古祠堂

周氏宗祠享堂 Xiangtang

周氏宗祠一进天井 Parvis of the first section

周氏宗祠二进天井　Parvis of the second section

棠樾鲍氏支祠 / 歙县棠樾村

棠樾鲍氏支祠位于歙县棠樾村牌坊群的西头。俗称男祠,又名万四公支祠,始建于明嘉靖末年,清嘉庆年间重建,祠堂坐北向南,占地约1000平方米,共三进五开间,通面阔15.98米,通进深47.11米。门厅原为五凤楼建筑,后毁坏,现仅剩八字墙及前檐石柱;二进廊庑也已无存,天井尚保存完整;大厅构筑宽敞,彻上露明造,抬梁与穿斗式构架相结合,前檐步做成船篷轩,其后做复水椽,敦本堂的细部做法反映了当时的地方特色,如梁头出挑承檐,斜撑辅助支撑,享堂后檐下做有斗栱,不承托檐部重量,起装饰作用;普通使用的梁头、雀替,梁头部分有做成象头状的,形象逼真,斜撑、雀替、平盘斗上雕有装饰花草,雕刻精美。寝殿台基很高,前檐立有青石板栏杆,梁架形式同中庭,在其后部设有须弥座式牌位座;前檐两侧墙上各竖有"义田禁碑"一块;后进天井两庑墙上嵌有"鲍氏义田记"碑刻八块。宋以后,凡历代鲍氏以孝行著名者,均奉祀于该祠。

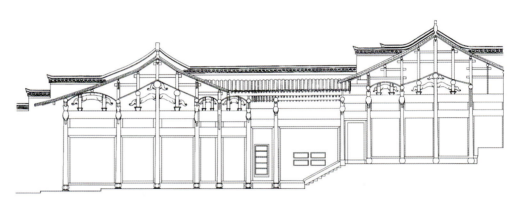

棠樾鲍氏支祠剖面图 Section view

Branch Ancestral Hall of the Bao Clan in Tangyue

Tangyue Village, Shexian County

Branch Ancestral Hall of the Bao Clan is located west of the Group of Gateways in Tangyue Village, Shexian County, with another name of Branch Ancestral Hall of Wan Sigong and the local name of Nan Ci, the hall for male ancestors. It was first built at the end of the Jiajing reign of the Ming Dynasty and rebuilt in the Jiaqing reign of the Qing Dynasty. Facing south, it covers an area of 1000 square meters and has five-bayed three sections altogether. It is 15.98 meters in width and 47.11 meters in depth. The gate hall is of the style of five-phoenix. It was destroyed afterwards, and only the splay screen walls and the front stone columns are left. Nothing remained of the corridors of the second section, but the parvis is kept almost untouched. The structure of the main hall is spacious and constructed with exposed roof frame without ceilings. It combines the post and lintel construction and the column and tie construction, with the sail-shape front eaves. The details of Dunben Hall show the features of local constructions. For instance, the head of beam stretches to hold the eaves; the batter brackets assist the bracing; the dougong, or bracket set, below the rear eaves of Xiangtang only decorates in stead of holds the weight of the eaves; the sparrow brace and head of beam are in common use; the head of beam is made lifelike head of elephant and the brackets, sparrow braces and pingpandou are decorated with fine carvings of flowers and grass. The platform of Qindian is very high with black stone railing panels below the front eaves. Its beam structure is the same as the atrium. At its back are placed the xumizuo-based tablets. Two steles of "Yi Tian Jin Bei", which means what was not allowed in the public fields, stand on both sidewalls under the front eaves. There are eight steles on which "Annals of the Public Fields" was inscribed on both side walls along the corridors in the skywell of the last section. All the Baos who were known for their filial behaviors were worshiped here since the Song Dynasty.

棠樾鲍氏支祠全貌 A full view

棠樾鲍氏支祠立面及广场 Facade of the ancestral hall and the square

棠樾鮑氏支祠

棠樾鲍氏支祠敦本堂 Dunben Hall

棠樾鲍氏支祠一进大厅 Main hall of the first section

棠樾鲍氏支祠侧立面图 Side elevation

棠樾清懿堂

歙县棠樾村

棠樾清懿堂位于歙县棠樾村牌坊群的西头。俗称女祠，始建于清嘉庆年间，祠堂坐南朝北，占地约1000平方米，共三进五开间，该祠大门外建有围墙，围墙中部有一长方形漏花窗，其入口开在东北角上，其门宽不足1.5米，门上有"清懿祠"砖雕匾额，门厅深五檩用三柱，两边檐柱为青石质，中柱原装门并立有抱鼓石，现存两边八字花墙，上有精致砖雕。中庭作拜殿，全堂彻上露明造，深九檩，明间后金柱间原装照壁，前后檐步做成船篷轩，檐之间做复水椽，梁架除山面为穿斗式外皆为抬梁式。寝堂为奉祀祖先牌位场所，地坪高起，内设须弥座式牌位座，深亦九檩，构架同中庭，不同之处是前檐檐下安有带枫栱的斗栱，以作装饰。该祠的天井铺地、檐柱、栏板均为青石，极其规整。据《歙县志·义行》载，鲍启运因"家祠旧奉男主，未祀女主，遗命其子重建女祠"。该祠作为专奉女主的祠堂实属罕见。

棠樾清懿堂入口 Entrance

Qingyi Ancestral Hall in Tangyue

Tangyue Village
Shexian County

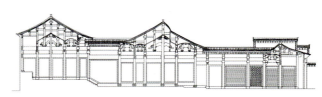

棠樾清懿堂剖面图 Section view

Qingyi Ancestral Hall, having a local name of Nu Ci, the hall for female ancestors, is located to the west of the Tangyue Group of Gateways, Shexian County. It was first constructed in the Jiaqing reign of the Qing Dynasty. Facing south, it covers an area of 1000 square meters and is five-bayed and three-sectioned. There are bounding walls outside its gate. In the middle of the wall there is a rectangle ornamental fretted window. The door of the bounding wall, less than 1.5 meters wide, is at the northeast corner with a brick-carving board of "Qingyi Hall" above it. The gate hall is five purlins deep, with three columns. The side eave columns are made of black stone. There used to be a door at the newel with drum-shaped bearing stones, but now only the splayed screen walls are left, decorated with delicate brick-carvings. The middle hall is the hall of worship, constructed with open rafters, nine purlins in depth. There used to be a screen wall between the principal columns behind the central bay. The eaves are sail-shaped, and the beams are mainly post and lintel construction except some of the column and tie construction. Qindian, the place for the memorial tablets of the ancestors, is nine purlins in depth, with the level ground lifted and the xumizuo-based tablets in. The structure is the same as the middle hall, but there is a decorated dougong with fenggong below the front eaves. The skywell is paved with and all the eave columns and railing panels are made of black stones. According to the Annals of Yixian County, Bao Qiyun "said in his last will that an ancestral hall for females must be constructed because the family hall was only for masters and nowhere was for mistresses." This hall is unusual, for only the matriarchs were worshiped here.

棠樾清懿堂正堂 The main hall

棠樾清懿堂后进天井 Parvis

棠樾清懿堂天井 Parvis

棠樾清懿堂前院 The front courtyard

棠樾清懿堂入口一隅 A corner of the entrance

郑村郑氏宗祠

歙县郑村

郑村郑氏宗祠坐落于歙县城西的郑村，建于明万历四十三年（1615年），祠堂坐北朝南，沿中轴线上布置门坊、门厅、廊屋、享堂、寝堂、天井等。祠堂前牌坊为青石仿木结构，建于明代末年（1615年），四柱三间五楼，高大雄伟，柱、梁、枋上遍饰包袱锦彩绘纹饰，典雅工整。门厅作悬山式，檐下斗栱装有枫栱，穿过门厅，两侧有廊屋。享堂构筑宏伟，彻上露明造，梁架作叠梁式，前檐步做轩，屋下增做一层复水椽，梁柱用材硕大，月梁、瓜柱、平盘斗均施云头卷草雕饰，外檐外间与角柱间铺作斗栱华美，形制古朴，堂上悬有"济美堂"、"道义传宗"匾，寝堂进深较浅而高度不减，内设须弥座式牌位座。寝堂台基外沿立石栏，栏版、望柱等处雕刻精美细致。

郑氏宗祠门厅 The gate hall

Ancestral Hall of the Zheng Clan in Zheng Village

Zheng Village of Yixian

Ancestral Hall of the Zheng Clan is located in Zheng Village in the west of Yixian County and was constructed in the forty-third year of the Wanli reign of the Ming Dynasty (1615). Facing south, along the central axis are the gate arch, the veranda, the lobby, Xiangtang, Qindian and the parvis. The memorial gateway in front of the hall is black stone imitation of the wood-frame construction, which was set up at the end of the Ming Dynasty (1615). It is grand and majestic with four pillars, three bays and five towers. The pillars, the beam and the purlin tiebeams are all decorated with coloured drawings, graceful and orderly. The lobby has the xuanshan--Chinese overhanging gable roof. The dougong under the eaves are installed with fenggongs. There are verandas on two sides behind the lobby. The structure of Xiangtang is magnificent, and all is constructed with open rafters. The beam frame is laminated and the beam-columns are huge; the crescent beam, short columns and pingpandou are all decorated with the sculptures of wavy crocket cloud pattern. The dougong between the outer-eaves and the corner columns is gorgeous and its shape and structure are simple and unsophisticated. In the hall are hung the boards of "Ji Mei Hall" and "Dao Yi Chuan Zong", which means morality and justice carried on through generations. Qindian is shadow in depth but has the same height and in it are the xumizuo-based tablets. Outside the platform of Qindian stand stone railings, with delicate and exquisite inscriptions on the railing panels and the baluster shafts.

郑村郑氏宗祠

郑氏宗祠入口处牌坊 Gateway at the entrance

中国徽派建筑　古祠堂

郑氏宗祠一进　The first section

郑氏宗祠大门 Entrance

郑氏宗祠梁架 Beam construction

郑氏宗祠二进 The second section

司谏第 / 歙县潜口村

司谏第位于歙县潜口村，始建于明弘治八年（1495年），砖木结构，二进三间，系明永乐初进士、吏科给事中汪善的五个孙子为祭祖所建的家祠，宅第门前的"司谏第"坊额为明代原物。明间阑额上有"劲节高标"四字匾额。享堂中悬明成祖敕谕匾额一方。另有《奉政大夫江公祠堂记》碑刻。碑记云："此非汪氏通族之祠也，一家之祠也"。现存享堂三间，通面阔8.2米，进深8米。司谏第用材硕大、梭柱、月梁、荷花墩、叉手、步梁及斗栱均有精美的雕刻，特别是枫栱的雕刻宛如飞卷的流云，尽显一代建筑风尚。上昂铺作，更是罕见，整个厅堂充分体现了明代早期的建筑风格。

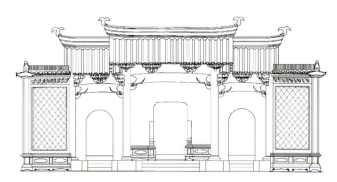

歙县潜口司谏第正立面图 Facade of Sijiandi

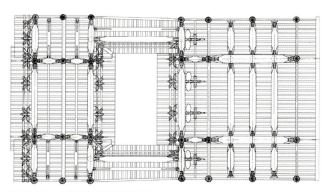

歙县潜口司谏第仰视平面图 Upward plan of Sijiandi

Sijiandi – Residence of the Remonstrator

Qiankou Village, Shexian County

Residence of the Remonstrator is located in Qiankou Village, Shexian County. It was first built in brick-and-timber construction, three-bayed and two sections, in the eighth year of the Hongzhi reign of the Ming Dynasty (1495). It was the family memorial hall that was set up by five grandsons of Wang Shan, a Jinshi in the early Yongle reign of the Ming Dynasty and supervising censor in Office of Scrutiny for Personnel. The architrave of "Residence of the Remonstrator" hung above the gate is the original one of the Ming Dynasty. Four characters of "Jin Jie Gao Biao" are written on the board of the central bay. A board conferred by Chengzu, an emperor of the Ming Dynasty, is hung in the center of Xiangtang, the main hall. There is also a stele of Annals of the Grand Master for Governance Mr. Jiang, saying "this is not the memorial hall for the whole clan but only for the family". Only a three-bayed Xiangtang is left, 8.2 meters in width and 8 meters in depth. The materials used for the construction are gigantic. The fine carvings on the shuttle-shaped columns, the crescent beam, the lotus pillars, the inverted V braces, step-cross beams and dougong, especially the carvings of the swiftly floating clouds on the fenggong reflect the architecture fashion of that time. The expense was unusually high and the whole hall fully embodies the architecture style of the early Ming Dynasty.

司谏第正门 Entrance

司谏第内景 Internal view of Sijiandi

三槐堂

休宁县溪头村

三槐堂位于休宁县溪头村，又称王家大厅，系明代举人王经天故宅。原为三进，现后进已毁，仅存门屋、享堂二进，占地面积约1000平方米。砖木结构，有柱182根，主柱围粗1.4米。站在门厅，放眼望去，百柱耸立，气势磅礴，柱上支撑雕镂平盘斗，下垫刻花柱托和石雕柱磉。梁架作冬瓜形月梁，梁下用雕花雀替承托梁头，檐口斗栱均为五踩，柱头科为插栱，平身科为十字形斗栱。前、中两进之间开大天井，两侧配厅又各有小天井。总体结构严谨，井然有序，享堂檐下悬有"龙章宠锡"的匾额，现闭目回想，"三槐堂"在鼎盛时竟是何等的辉煌，难免有民间"金銮殿"之称。

三槐堂正门 Entrance

Sanhuai Hall–Hall of Three Chinese Scholartrees

Xitou Village, Xiuning County

Sanhuai Hall is located in Xitou Village, Xiuning County, also named the Hall of the Wangs. It is the former residence of Wang Jingtian, a provincial graduate of the Ming Dynasty. There used to be three sections, but the last section was damaged and only the gate hall and Xiangtang remained. It covers an area of 1000 square meters, a brick-and-timber construction with 182 columns of which the main column is 1.4 meters in girth. Seen from the lobby, hundreds of columns stand with great momentum. On top of the columns are the chased pingpandous, and at the bottoms are the column plinths engraved with flowers and carved stone column pedestals. The beam frame is a crescent beam and under it are the engraved sparrow braces supporting the heads of the beam. There is a large courtyard between the front and the middle sections and two small skywells in the wing halls on both sides. The overall arrangement is well knit and in good order. The board of "Longzhang Chongxi" is hung below the eaves of Xiangtang. Imagine how magnificent Sanhuai Hall was in its old days, hence it got the name of "Local Throne Hall" among the natives.

三槐堂的匾额"龙章宠锡"　Inscribed board of "Longzhang Chongxi"

三槐堂天井　Parvis

三槐堂天井享殿 Parvis and Xiangtang

三槐堂

进士第 / 休宁县黄村

进士第位于休宁县黄村，建于明代嘉靖年间，系嘉靖进士黄这福所建。规模宏伟，气势壮观的大门门楼有七层斗栱。脊高12米，进深51米，前后共四进，依次为门楼、门屋、享堂、寝楼，每进庭院两侧均有侧廊相连。但门屋前增建有门楼，寝楼后又加一天井，并在后天井的垣墙上做假门楼，在纵轴线上形成进深为五进的格局。占地面积790平方米。该房门楼上嵌有木匾一块，上书"进士第"三个大字。整座建筑有木柱102根，主柱围粗1.6米，选料讲究。横梁上雕镂龙、凤、狮、虎等异禽猛兽，刀法细腻，形象生动。

进士第柱及柱础 Columns and their bases

Jinshidi
—Residence of Jinshi (a successful candidate in the imperial examination)

Huang Village, Xiuning County

Jinshidi is located in Huang Village, Xiuning County. It was constructed by Huang Zhefu, a Jinshi of the Jiajing reign in the Ming Dynasty. The grand and majestic gate tower has seven levels of dougong. The residence is 12 meters in height and 51 meters in depth. There are four sections altogether— gate tower, gate hall, Xiangtang and Qindian. Side corridors lie between sections. There is an additional gate tower in front of the gate hall and a parvis is added behind the Qindian. On the wall of the rear parvis is an artificial gate tower. So the structure of five sections is arranged along the longitudinal axis. The residence covers an area of 790 square meters. On the gate tower is a wooden board with characters of "Jin Shi Di" on it. There are 102 wooden pillars in the whole construction and the main columns measure 1.6 meters in perimeter. The materials were selectively chosen. On the crossbeam is finely engraved lifelike animals of dragons, phoenixes, lions and tigers.

进士第一进天井 Parvis of the first section

进士第二进天井 Parvis of the second section

进士第正门 Entrance

进士第

经义堂

婺源县黄村

经义堂位于婺源县黄村，系黄氏宗祠，又名百柱宗祠（该祠共有杉木柱100余根），建于清朝康熙年间。背山面水，坐北朝南，祠堂为砖木结构，由前院、门楼、享堂、寝殿组成，面积1200平方米。前院两侧耳门相通，内列乾隆年间八棱旗杆石四对。前进为九脊顶，五凤楼式门楼。大门用上等杉木制成，两边分刻八仙，惟妙惟肖。穿过门楼，是享堂天井，天井用青石铺成，平整光洁。正堂中央悬挂清朝文华殿大学士张玉书手书"经义堂"的匾额，遒劲有力，为一代大师墨宝。享堂柱共三排，每排四根，柱身粗大，且柱础各异，各具风格，雕刻粗犷有力。大梁上雕刻精致，四个石基雕刻有"鹭鸶戏莲"、"凤戏牡丹"、"仙鹤登云"及"喜鹊含梅"。前行走过享堂，在寝殿和享堂之间有一道门楼，这便是"九级金阶"，现仅存七级。据《黄氏宗谱》记载，该祠初建时，第三层原有台阶九级，因避讳皇宫"九级金阶"，故撤掉两级。因依山而建，寝殿高于前堂1.5米，寝殿气势雄伟，高达十余米，其内雕刻朴实无华，大处苍劲有力，细部惟妙惟肖，整个"经义堂"完全是一本不可多得的建筑史书。

Jingyi Hall

Huangcun Village, Wuyuan County

Jingyi Hall is located in Huangcun Village, Wuyuan County. It is the ancestral hall for the clan of Huangs and also named the Ancestral Hall of Hundred Pillars, for it has more than 100 columns of fir timber. It was set up in the Kangxi reign in the Qing Dynasty. Facing south towards the water, it is a brick-and-timber construction and consists of the front yard, gate tower, Xiangtang and Qindian, covering an area of 1200 square meters. The front section is a five-phoenix style gate tower. The gate is made of first-class fir and the Eight Immortals in the legend are carved on it remarkably true to life. Next to the gate tower is the courtyard smoothly paved of black stones. The board of "Jingyi Hall" is hung in the middle of the main hall, written by Zhang Yushu, a famous scholar in the Qing Dynasty. The calligraphy is vigorous in strokes and it is a treasured piece of the outstanding masters of that time. There are four rows of columns in the main hall and four columns for each row. The columns are thick and their bases are different in style. The carvings are rough and forceful. The engravings on the beams are delicate and the four stone foundations are engraved with the designs of Egrets on Lotuses, Phoenix and Peony, Crane on Clouds and Magpie with Plum Blossom. Next to the main hall, there is a gate tower between Qindian and Xiangtang. That is the Nine Gold Steps, but only seven steps are left. According to Annals of the Huang Clan, there used to be nine steps on the third level when it was first constructed, but two steps were removed to avoid the taboo of the Nine Gold Steps in the imperial palace. Qindian, built on the slope, is 1.5 meters higher than Xiangtang, with a height of more than ten meters. The carvings inside are simple and vigorous but the details are absolutely lifelike. The whole Jingyi Hall is a rare book of history of architecture.

经义堂寝殿入口"九级金阶"　　Nine gold steps at the entryway of Qindian

经义堂享堂 Xiangtang

中国徽派建筑　古祠堂

经义堂梁架　Beams

经义堂卷棚　Round ridge roof

经义堂前檐童柱　Double columns

经义堂天井　Parvis

经义堂入口　Entrance

经义堂柱础 Column bases

   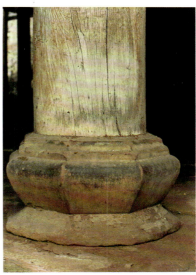

经义堂柱础 Column bases

经义堂柱础 Column bases

俞氏宗祠

婺源县汪口村

俞氏宗祠位于婺源县东北28公里的汪口村。建于清乾隆年间，占地面积700平方米。中轴对称，坐西北朝东南，平面呈长方形，宽15.6米，纵深42.6米，周环高10米的砖墙，祠堂为三进院落。门楼为木结构五凤楼，歇山顶，青瓦覆盖，戗角高翘。门楼正面，檐下斗栱密布，横枋刻双龙戏珠图案，横枋下面明枋深雕双凤朝阳。门楼内里，前间顶部用木板卷棚，后间为平天花。由两廊与游亭达正厅，横梁直接搭接巨大的石柱，左右作吊柱支撑的垂柱上端，精雕雌雄狮子相对，栩栩如生。祠堂中进三间，前、后进各五间，均有天井，共有柱70根，地面、天池、台阶全铺青石板。前、后进走廊两侧有小圆门通花园，花园内遍植奇花异草，另有百年木樨三棵。整个祠堂以细腻的雕刻工艺见长，凡梁枋、斗栱、脊吻、檐椽、驼峰、雀替等处均巧琢雕饰，有浅雕、深雕、圆雕、透雕形式的龙凤麒麟、松鹤柏鹿、水榭楼台、人物戏文、飞禽走兽、兰草花卉等精美图案百余组，被誉为木雕艺术的宝库、建筑艺术的殿堂。

俞氏宗祠正门 Entrance

Ancestral Hall of the Yu Clan

Wangkou Village in Wuyuan County

Ancestral Hall of the Yu Clan is located 28 kilometers northeast to Wangkou Village in Wuyuan County. It was constructed in the Qianlong reign of the Qing Dynasty, covering an area of 700 square meters. Facing southeast, it is rectangle (15.6 meters by 42.6 meters), and symmetrical by the central axis, with a ten-meter-high brick wall surrounding it. The hall has three sections. The gate tower is in a five-phoenix structure, the xieshan roof covered by gray tiles and the diagonal ridges soaring high. The dougongs gather under the front eaves of the gate tower and the designs of dragons and phoenixes are engraved on the architraves. Inside the top of the gate tower is the vaulted roof of planks in the front part and flat ceiling in the rear part. The two corridors and the pavilion lead to the main hall. The crossbeam overlaps directly on the huge stone columns. Lifelike sculptures of lions are carved on the upper part of the drooping columns. The middle section is three-bayed while the front and rear sections are five-bayed with a parvis for each section. There are 70 columns altogether. The ground, skywells and the steps are paved with gray stone boards. In the corridors of the front and the rear sections there are moon gates leading to the garden where exotic flowers and rare herbs are planted, as well as three sweet-scented osmanthuses trees of hundreds of years. The hall is famous for its exquisite carvings. All the beams, architraves, dougongs, jiwens, eave rafters, sparrow braces and the camel-hump shaped supports are skillfully carved and delicately engraved; there are low-relieves, high-relieves, medallions and openwork carvings of hundreds of beautiful designs of dragons and phoenixes, kylins, cranes and deer, pines and cypresses, waterside pavilions and towers, birds and beasts, fragrant thoroughwort and plants. It is known as treasury of woodcarvings and thesaurus of architecture.

俞氏宗祠入口檐下斗栱 Bracket sets under the eaves of the entrance

俞氏宗祠四水妆堂天井 Parvis

中国徽派建筑　古祠堂

俞氏宗祠入口背面　Back of the entrance

俞氏宗祠

俞氏宗祠正门檐口斗栱 Dougong at eaves of the entrance

俞氏宗祠门匾 Inscribed board above the gate

俞氏宗祠天井屋檐 Eaves of the parvis

俞氏宗祠二进天井　Parvis of the second section

俞氏宗祠梁枋木雕　Wood carvings on the beams

俞氏宗祠梁架　Beams

俞氏宗祠享堂顶棚　Roof of Xiangtang

俞氏宗祠享堂二层　Second floor of Xiangtang

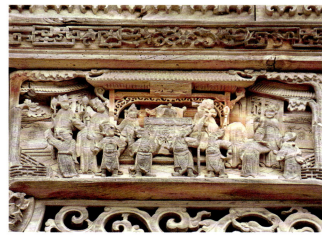
俞氏宗祠檐下木雕 Wood carvings under the eaves

俞氏宗祠梁上木雕 Wood carvings on the beams

俞氏宗祠雀替 Sparrow brace

俞氏宗祠前檐梁枋 Tiebeam of the front eaves

俞氏宗祠丁头栱 T-shaped arm

俞氏宗祠梁枋细部 Details of the tiebeams

俞氏宗祠镂空雕平盘斗
Pierced-carved pingpanmen

俞氏宗祠象鼻及平盘斗
Xiangbi or elephant's nose and pingpanmen

成义堂

婺源县龙山乡

成义堂位于婺源龙山乡，又名豸峰堂，建于清同治年间，成义堂属六家房支祠中一个房头的祠堂，当时六家后裔中有兴来公一脉，家境富足，出钱兴建了成义堂。成义堂前临桃溪水，其形制颇有特点。祠平面基本保持中轴对称，入口两侧建有八字照壁，门屋为顺应沿河地势，与主轴线偏转了一个角度，进入门屋后的第一进庭院由四面高墙围合成一个封闭空间，院内没有廊庑，两侧有小门可通附院，正面墙上用砖雕砌成三开间牌楼门形式，上刻"通奉大夫晋三公祠"。穿此入内为一"四水归堂"的庭院，三面围廊，后为两层高的寝殿。与一般祠堂不同，成义堂内没有享堂，而改为扩大的廊间，祀祭时原应放在享堂内进行的活动却移到了寝殿楼下。寝殿是整座建筑最为华丽的部分，仅斗栱就有十字形、米字形、斜栱等近十种。梁垫、柱础等处多作高浮雕、透雕的木雕。最为精彩的是寝殿底层，正面月梁两端的雀替是倒爬着的两只木雕狮子，神态惟妙惟肖。大厅天花正中有一螺旋形斗栱构成的覆钵藻井，堪称徽派建筑中的一绝。置身"成义堂"中，你不得不为古人精湛的建筑艺术所叹服。

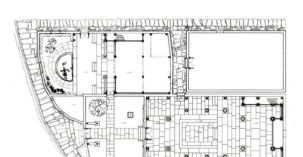

成义堂底层平面图 Plan of the first floor

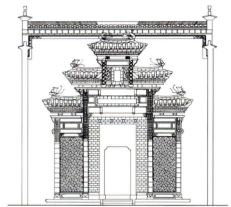

成义堂第一进院落砖雕门楼立面图
Facade of the brick-carved entry hall of the first courtyard

Chengyi Hall

Longshan Village of Wuyuan County

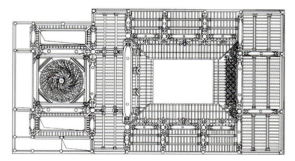

成义堂仰视平面图 Upward plan of Chengyi Hall

Chengyi Hall is located in Longshan Village of Wuyuan County, and also named Zhifeng Hall. It is constructed in the Tongzhi reign of the Qing Dynasty. Chengyi Hall stands facing the Taoxi Brook and is peculiar in its shape and structure. The hall is symmetrical by the central axis and at the entrance stands the splay screen wall. The gate hall deflects a little angle from the principal axis along the brook. The first section behind the gate hall is a closed space surrounded by four high walls and there are no corridors in it. Two small doors on both sides lead to the adjacent yards. On the front wall is the brick-carved three-bayed gate tower with an inscribed board. Through it is a courtyard, which is called "Rain water collected in the central courtyard". Corridors stand on three sides of the courtyard and the other is a two-storied Qindian. Different from other ancestral halls, there is no Xiangtang in Chenyi Hall but an extended veranda instead. The rites would be held on the first floor of Qindian instead of Xiangtang. Qindian is the most gorgeous part of the construction. There are about ten kinds of dougongs in it – the cross shaped, □-shaped, and the skew arch. The beam pads and the column bases are usually carved in high-relieves and openwork carvings. The most wonderful carvings are in the first floor of Qindian. On the sparrow braces at both ends of the crescent beam are woodcarvings of two lifelike lions hung upside down. The overturned caisson on the ceiling of the hall is made of spiral dougong, which is unique in the architecture of Huizhou. Everyone would admire the art of ancient architecture when standing in Chengyi Hall.

成义堂天井俯视 Perspective view of the parvis

成义堂寝殿中的藻井 Caisson ceiling of Qindian

中国徽派建筑　古祠堂

成义堂二层远眺 Distant view from the second floor

成义堂一进院落 Courtyard of the first section

成义堂

成义堂庭院　Courtyard of Chengyi Hall

成义堂檐廊下的木雕 Wood carvings under the eaves

成义堂入口倒座梁枋雕刻 Carvings on the beams of the entry hall

成义堂

成义堂轩廊雕刻 Carvings of the side gallery

成义堂轩廊斜撑 Brackets of the side gallery

成义堂檐口雕刻 Carvings of the eave edging

成义堂入口抱鼓石 Drum-shaped bearing stones at the entryway

成义堂入口"簪缨世家" Characters above the main gate

成义堂院落 Courtyard of Chengyi Hall

63

龙川胡氏宗祠

绩溪县大坑口村

Ancestral Hall of the Hu Clan in Longchuan

Dakengkou Village in Jixi County

　　龙川胡氏宗祠坐落于绩溪大坑口村,宗祠始建于宋,明嘉靖年间大修并扩建,清光绪二十四年再度大修。建筑仍保持了明代徽派建筑艺术风格,祠堂坐北朝南,由照壁、平台、门楼、庭院、廊庑、祭堂、厢房、寝室、特祭祠等部分组成,面积为1146平方米,且宗祠由前至后依次递增高度,中轴对称布局,合理严谨,蔚为壮观,堂前广场面对是宽近20米的大型八字照壁,一条小溪从祠堂和照壁之间穿过,门楼为重檐歇山式,布局匀称。仪门上彩绘尉迟恭、秦叔宝两门神,仪门的上方悬挂着"江南第一祠"的匾额,檐下方梁梁面雕刻图案精美,真可谓巧夺天工。门楼后面为天井。20根石柱同20根月梁衔接,排列在天井四周,擎起东西两廊和前中两进的南北房檐。过天井是中进,为祠堂正厅。由14根围粗达166厘米的银杏树圆柱架着大小19根冬瓜梁构成。正厅两侧各为高达丈余的落地窗门,每扇窗上截有镂空花格,下截是平板花雕。祠堂内所有的柱础均为石础,风格各异。正厅上首也是一排落地窗门,花雕画面以鹿为中心,衬以山光水色,竹木花草。穿过正厅的右侧的一小门,是特祭祠,特祭祠布局紧凑,空间设计巧妙。穿过前厅,后进是寝殿,寝殿高于前堂,分上下两档,中隔一个狭长的天井。寝室窗门雕刻的全是花瓶,采用浮雕和浅刻着八仙道具、文房四宝、书案画卷、圆椅条桌等,再现出匠师们以刀代笔、驰骋艺林的高超技术和创作天赋。

龙川胡氏宗祠所在大坑口村全景　Overall view of Dakengkou Village

Ancestral Hall of the Hu Clan is located in Dakengkou Village in Jixi County. It was first constructed in the Song Dynasty and in the Jiajing reign of the Ming Dynasty it was rebuilt and expanded. The structure keeps the artistic style of Huizhou architecture of the Ming Dynasty. Facing south, it consists of screen wall, terrace, gate tower, courtyard, corridors, sacrificial hall, wing rooms, Qindian and the hall for special sacrifice. It covers an area of 1146 square meters. The hall grows higher and higher progressively from the front to the back. It is symmetrical by the central axis and it is rationally arranged and compactly laid out. The huge splay screen wall, which is almost 20 meters wide, stands in the square in front of the hall; and a brook runs across between the hall and the screen wall. The gate tower has the multiple-eave xieshan roof in proportional arrangement. On the Yi Gate, there are coloured drawings of two door gods – Yuchi Gong and Qin Shubao, and above the gate hangs the inscribed board of "The First Hall in South of the Yangtze River". The surface of the square beams under the eaves are decorated with delicate carvings of beautiful designs. Behind the gate is a parvis. Standing around the parvis, twenty stone columns are connected to twenty crescent beams. They hold the east and west corridors and the north and south eaves of the front and rear rows of houses. Next to the skywell is the middle section, which is the main hall. There 14 columns made of ginkgo tree, which measure 166 centimeters in perimeter, support 19 oval beams. On both sides of the main hall are down-to-floor partition doors with pierced carvings on the upper part and plate carvings on the lower part. All the column bases in the hall are made of stone and different in style and shape. The front of the main hall also consists of a row of partition doors and their engravings are mainly design of deer. A small door on the right of the main hall leads to the hall for special sacrifice. It is compactly laid out and ingeniously planned. Through the front hall is the last section—Qindian. Qindian is higher than the front hall and separated into two parts by a long and narrow skywell. The windows and doors of Qindian are engraved in vases, and in low relieves of Eight Immortals' tools, four treasures of the Chinese study, writing desks with scrolls, and so on. All these reflect the craftsmen's superb skill and creative gift of carving.

中国徽派建筑　古祠堂

龙川胡氏宗祠正面全貌　A full view of the front of the Ancestral Hall

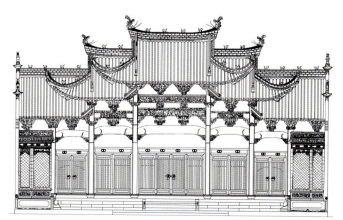

龙川胡氏宗祠正立面图　Facade of the Ancestral Hall

龙川胡氏宗祠

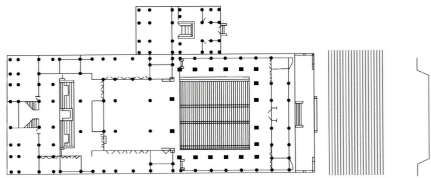

龙川胡氏宗祠底层平面图 Plan of the first floor

龙川胡氏宗祠祭堂及天井侧座 Sacrificial hall and the side gallery of the yard

龙川胡氏宗祠

中国徽派建筑　古祠堂

龙川胡氏宗祠祭堂前檐封板雕刻　Carvings on the eave board of Sacrificial hall

龙川胡氏宗祠正门　The main gate

龙川胡氏宗祠

龙川胡氏宗祠寝堂前天井 Parvis in front of Qintang

龙川胡氏宗祠特祭祠 Ancestral hall for special sacrifice

中国徽派建筑　古祠堂

龙川胡氏宗祠二层远眺　Distant view from the second floor

龙川胡氏宗祠寝堂及天井　Qintang and the parvis

龙川胡氏宗祠祭堂及天井　Sacrificial hall and the courtyard

龙川胡氏宗祠檐下木雕　Wood carvings under the eaves

龙川胡氏宗祠隔扇门木雕　Wood carvings on the partition doors

龙川胡氏宗祠裙板木雕　Carvings on the panels

龙川胡氏宗祠

龙川胡氏宗祠梁脐灯钩 Lantern hook on the beam

龙川胡氏宗祠梁脐灯钩(云龙吐珠) Lantern hook on the beam

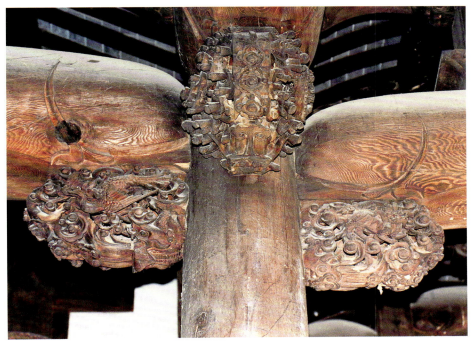
龙川胡氏宗祠梁头及雀替雕刻 Carvings on the beam head and the sparrow brace

贞一堂

祁门县渚口村

贞一堂位于祁门县渚口村。始建于明，1673年重建，1910年再建。坐北朝南，占地面积1267平方米。砖木结构，分前、中、后三进。大门两侧有黟县青石鼓一对，雕饰有龙凤呈祥、麒麟送子等图案。进门为前进，左右两厢，中为通道。过天井即为正厅，正厅有108根立柱支撑，规模宏大，堂中高悬"贞一堂"的匾额，是春秋二季祭祀和举行庆典的场所。后进有东西天池各一，天池四周石柱和栏板上，均刻有各式各样的花鸟图案，如鲤鱼喷月、雁落荷花、松鹤延年、松柏常青等，雕刻精细。过天池为享堂，是供奉渚口倪氏祖宗神位之地。

贞一堂正厅前天井 Parvis of the main hall

Zhenyi Hall

Zhukou Village in Qimen County

Zhenyi Hall is located in Zhukou Village in Qimen County. It was first constructed in the Ming Dynasty and rebuilt in 1673 and 1910. Facing south, it covers an area of 1267 square meters. It is the brick-and-timber construction and consists of three sections. A pair of stone drums of Yixian Black, decorated with inscriptions of dragon and phoenix, stand beside the gate. Through the gate is the first section, which consist of two wing rooms and a passage in the middle. Next to the parvis is the main hall, where sacrificial ceremonies in spring and autumn are carried out and the celebrations are staged. It is imposing that the main hall is held by 108 columns. In the center is hung an inscribed board of "Zhenyi Hall". There are two ponds in the rear section. The railing panels around the two ponds are carved with various designs of flowers and birds. Next to the ponds is Xiangtang--the main hall, where the memorial tablets of the ancestors of Ni Clan in Zhukou are placed.

贞一堂享堂 Xiangtang

贞一堂入口 Entryway of Zhenyi Ancestral Hall

中国徽派建筑　古祠堂

贞一堂二进天井 Parvis of the second section

　　Yuqing Hall is located in Zhulin Village in the town of Xin'an, Qimen County. The ancient stage of "Yuqing Hall" is also named "Longxi Heavenly Water Eternal Stage", and "Eternal Stage" in abbreviation. It makes up the first section of the ancestral hall of the Zhao clan and it was set up in the early years of the Xianfeng reign (1851-1853 A.D.). The ancestral hall consists of three sections. Now only an area of 500 square meters is remained and the first section is the stage. Facing west, the stage sits opposite to the main hall. The wing rooms are used as galleries for watching plays and are connected to the main stage. It is a richly ornamented building, resplendent and magnificent and tasteful in architecture technology. Over the center of the front stage is the coloured caisson ceiling with streamline settings inside for the purpose of performance. The lower opening is square with triangle plaques at four corners, on which the designs of two bats are inscribed. The architraves of front eaves, diagonal braces, queti and crescent beams are decorated with carvings of roles in plays and flower-and-bird designs. They are vividly shaped and the characters are lively. Spaces on both sides of the stage are used for the accompaniment of the band. There are fancy inscriptions of characters and flowers and birds in the galleries on both sides and every compartment has delicate partition windows. There are side doors leading outward in both corridors under the wing rooms. The stage sits opposite to the parvis and Xiangtang. There the clansmen watched the performance. The inscribed board of "Yuqing Hall" in Xiangtang can be seen horizontally on the stage. Yuqing Hall is an ancient stage that is preserved well throughout the country.

馀庆堂聚声的顶棚藻　Sound-gathering vaulted ceiling

余庆堂

藻井 Vaulted ceiling

中国徽派建筑　古祠堂

馀庆堂正厅　The main hall

馀庆堂右侧包厢看台（由内往外看）
View of the stage from the left-side gallery

馀庆堂左侧包厢看台（由外往内看）The right-side gallery

馀庆堂戏台正堂 The main hall

叶奎光堂

黟县南屏村

叶奎光堂位于黟县南屏村，为南屏叶氏支祠。建于明弘治年间，清雍正年间改建门楼和大门，乾隆年间重修祀堂及门楼，约700多平方米，祠堂门前有照壁，形成护垣。大门前有一广场，门两侧有一对雕刻精美的抱鼓石，门扇绘有一对门神，全祠分祀堂和享堂两大进，门楼高大，四根40厘米见方的石质檐柱，托着硕大的额枋和曲梁，上部为四柱三间三楼木质结构，明间三楼近10米宽的额枋上，开列着四攒九踩四翘品字斗栱。次间二楼各列二攒九踩四翘品字斗栱，结构相称的飞檐，相映生辉，穿过门楼，为祀堂和天井，天井尺度宏大，气势非凡，祀堂的正中悬挂着"奎光堂"的匾额，整个祀堂梁架没有过多的雕饰，朴实大方，享堂为二层楼，比祀堂面积略大。雕刻亦较祀堂精美。

叶奎光堂大门 Entrance

Ye Kuiguang Hall

Nanping Village in Yixian County

　　Ye Kuiguang Hall is located in Nanping Village in Yixian County and it is a branch ancestral hall of the Ye clan in Nanping. It was constructed in the Hongzhi reign of the Ming Dynasty and the gate tower and the main gate were rebuilt in the Yongzheng reign of the Qing Dynasty. In the Qianlong reign, the sacrificial hall and the gate tower were constructed again. It covers an area of about 700 square meters. The screen wall in front of the hall makes the wainscoting. There is a square in front of the entrance and besides the main gate are a pair of well-carved drum-shaped bearing stones. On the doors are the drawings of two door gods. The whole hall is divided into two parts—the sacrificial hall and Xiangtang or the main hall. The gate tower is very high and four stone peripheral columns, 40×40 centimeters in perimeter, hold the huge architraves and curved beams. The upper part of the gate tower is the construction of four columns, three bays and three towers. On the 10-meters wide architrave of the third tower, there are 品-shaped Dougongs. Next to the gate tower is the sacrificial hall and parvis. The parvis is uncommonly large in scale. Over the center of the sacrificial hall is hung the inscribed board of "Kuiguang Hall". The beam frames in the sacrificial hall are simple and tasteful with few sculptures. The two-storied Xiangtang is a bit larger than the sacrificial hall and the sculptures are more exquisite.

叶奎光堂享堂 Xiangtang

叶奎光堂天井倒座 the Parvis

叶奎光堂

叶奎光堂马头墙一角
Horse-head gable walls

叶奎光堂天井 Parvis

叶奎光堂

叶奎光堂入口
The entrance

叶奎光堂门前抱鼓石
Drum-shaped bearing stones at the entrance

敬爱堂

黟县西递村

敬爱堂位于黟县西递村中心，原为壬派胡十四世祖仕亨公住宅，始建于明万历二十八年，后毁于火，明末重建，清初落成。建筑面积达到1800余平方米。祠堂结构粗犷古朴，宏伟壮观。大门气势恢弘，进入正门前有一道木栅栏，由中门步入则为祭祀大厅。祭祀大厅是一大型天井，左右分设东西两庑，大厅分上下庭，上庭作为族祠的神圣场所，一直作为族事商议之堂，上庭之后为楼式建筑的供奉厅，供奉列祖列宗神位。祠堂祭祀堂正中悬挂"敬爱堂"的匾额，东西两庑梁檩间悬挂着的"天恩重沐"、"上国琳琅"、"四世承恩"、"盛朝英俊"四块金字匾，仿佛在向人们宣示胡氏族当年荣华、显赫。值得一提的是，"敬爱堂"后厅门头，高悬着一个斗大的"孝"，系南宋大理学家朱熹的手书，其字中有画、画中有字，细看字的上部，似一孝子仰面拱手跪地作揖，上部的反面则是牲畜（猴子）的嘴脸，暗寓是人须敬老爱幼，否则就是牲畜。故祠堂不仅是族人举办婚嫁喜庆的场所，也是教斥不肖子孙的场所。

南宋大理学家朱熹手书"孝"字
Character of "Xiao or filial piety" written by Zhu Xi

Jing'ai Hall
—Hall of Respect

Xidi Village, Yixian County

敬爱堂大门 Entrance

　　Jing'ai Hall is located in Xidi Village, Yixian County. Originally the residence of the fourteenth generation of the Hu clan and it was first set up in the twenty-eighth year of the Wanli reign of the Ming Dynasty. Later it was destroyed in fire, and rebuilt at the end of the Ming period and finished at the beginning of the Qing Dynasty. The constructions cover an area of about 1800 square meters. The structure of the hall is boorish and unsophisticated but grandiose. The gate is imposing and behind it is a row of wood fence. The middle door leads to the sacrificial hall which is a large yard with two corridors on both sides. The sacrificial hall is divided into two parts. The first part is the sacred place for the clan to hold ceremonies and to discuss the affairs of the clan. Behind the first part is the storied oblation hall for the memorial tablets of all the ancestors. The inscribed board of "Jing'ai Hall" is hung over the center of the sacrificial hall. Between the girders and the purlins of the corridors are hung the four gilded inscribed boards, which show the glory and brilliance of the Hu clan in those days. It deserves to be mentioned that the character of "Xiao—filial piety" hung on the door-head of the rear hall is hand written by Zhuxi, the famous Confucian in the Southern Song Dynasty. There are pictures in the character and the character is in pictures—the upper part of the character is like a dutiful son kneeling and making a bow submissively facing upward; but the other side of the character is a monkey's face. It means that a person should respect the old and love the young; otherwise he is no man but a beast. So the hall was the place not only for the clan to hold marriages and ceremonies, but also a place to teach lessons to unworthy descendants.

敬爱堂寝殿大堂 The main hall of Qindian

敬爱堂正面全貌　A full view of the hall's front

追慕堂(胡氏祠堂之支祠) Zhuimu Hall–Hall of Admiration (a branch of the Hu Clan Ancestral Hall)

敬爱堂享堂 Xiangtang

敬爱堂寝殿全貌 A full view of Qindian

敬爱堂

敬爱堂石柱 Stone columns

敬爱堂天井 Skywell

迪吉堂(胡氏祠堂之家祠) Diji Hall (a family ancestral hall of the Hu clan)

中国徽派建筑
HUIZHOU SCHOOL ARCHITECTURE OF CHINA

古牌坊

Ancient Gateways

许国石坊 / 歙县徽城内

许国石坊坐落于歙县县城阳和门东侧，又名大学士坊，俗称八脚牌楼，跨街而立，建于明万历十二年（1584年），是旌表"少保兼太子太保礼部尚书武英殿大学士许国"而立，牌楼平面呈长方形，南北长11.5米，东西宽6.77米，高11.5米。四面八柱，各联梁枋，整座牌坊由前后两座三间四柱三楼和左右两座单间双柱（前后两坊合用）三楼的石牌坊组合而成。全部采用于色茶园石。梁柱粗硕，方柱断面下大上小，且重心渐向坊心微偏，故结构安稳固实。石坊遍布雕饰，梁枋两端浅镌如意头，缠枝、锦地开光。中部菱形框内为深浮雕，如"巨龙飞腾"、"瑞鹤翔云"、"鱼跃龙门"、"威凤祥麟"、"龙庭舞鹰"、"三报喜"、"麟戏彩球"、"凤穿牡丹"等。直柱中段为散点团花式锦纹，上段为云纹，缀以姿态各异的翔鹤。柱基外侧的台基上，雕置蹲驻与奔走等各种动作的大狮子12只，有的大狮还抱弄小狮，形态生动活泼。台基左右侧皆镌各式瓣豸图案。石坊四面有"大学士"、"少保兼太子太保礼部尚书武英殿大学士许国"、"先学后臣"、"上台元老"等擘窠大字，出自明代书画家董其昌手笔。清人吴梅颠《竹枝词》云："八脚牌楼学士坊，题额字爱董其昌"。许国石坊的独特形制和建筑艺术，在全国是罕见的。

许国石坊上的精美石雕　Elaborate stone carvings on the gateway

Xu Guo Memorial Stone Gateway

Yanghe Gate of the town of Shexian County

Xu Guo Memorial Stone Gateway stands by the east side of Yanghe Gate of the town of Shexian County. It is also named "Gateway of Senior Scholars" and has a nickname of "Gateway with Eight Legs". Erected across the street, it was built in the twelfth year of the Wanli reign of the Ming Dynasty (1584), in honour of Xu Guo, a senior scholar and minister of the court. It is in a rectangular shape, 11.5 meters long from north to south, 6.77 meters wide from east to west and 11.5 meters high. Four sides and eight columns with architraves in between, the whole gateway consists of four gateways made of Chayuan Stone, two of which have three bays, four columns and three towers, and the other two have one bay, two columns and three towers. The structure looks solid and firm, with thick square beam columns. The stone panels are full of delicate traditional carvings. On the terrace by the column bases are carved twelve lively lions. There is an inscribed board on each of the four sides of the gateway, with characters written by Dong Qichang. The unique form and style of construction of this gateway is seldom seen in the country.

许国石坊夜景　Night view of the gateway

中国徽派建筑　古牌坊

许国石坊全貌　A full view of Xu Guo Gateway

许国石坊内侧　Internal view of the gateway

许国石坊恩荣正匾 The main inscribed board of "En Rong"

石坊上的精美石雕 Elaborate stone carvings on the gateway

石坊上的精美石雕 Elaborate stone carvings on the gateway

石坊上的精美石雕 Elaborate stone carvings on the gateway

许国石坊护坊石狮 The stone lion

许国石坊护坊石狮 The stone lion

同胞翰林坊 / 歙县唐模村

同胞翰林坊位于徽州区唐模村村口檀干园前山古道中间。建于清康熙年间，牌坊跨道而立，三间三楼，四柱冲天。通体采用茶园石雕筑而成，上雕飞禽走兽和各种图案，基座上有石狮四只，明间花板正反两面分别镌刻着"同胞翰林"、"圣朝都谏"，竖板上雕刻有"恩荣"二字，此坊为旌表唐模村许承宣、许承家兄弟而立，两人于康熙朝皆中进士，一授编修，一授庶吉士，均属翰林院，故有"同胞翰林"之称。

同胞翰林坊石坊正匾(正面) The main inscribed board (front)

同胞翰林坊石坊正匾(反面) The main inscribed board (back)

The Brother Hanlin (Members of the Imperial Academy) Gateway

Tangan Garden at the entrance of Tangmo Village

The Brother Hanlin Gateway is located in the middle of the ancient path in front of Tangan Garden at the entrance of Tangmo Village, Huizhou District. It was built in the Kangxi reign of the Qing Dynasty. Made of Chayuan Stone, the gateway stands across the road, three bays and three towers with four columns. There are various kinds of carvings on the gateway and its base is carved with four lions. The gateway was built in honour of the two brothers in Tangmo Village—Xu Chengxuan and Xu Chengjia, both of whom had passed the imperial examinations and become Jinshi in the reign of Kangxi, under the same Imperial Academy. Hence the title of the gateway.

同胞翰林坊

同胞翰林坊全貌　A full view of the gateway

棠樾牌坊群 / 歙县棠樾村

棠樾牌坊群位于歙县富揭乡棠樾村东大道上。共七座，明代建三座，清代建四座，"旌表"的对象和内涵可分为"忠、孝、节、义"几种类型，三座明坊为鲍灿坊、慈孝里坊、鲍象贤尚书坊。鲍灿坊旌表明弘治年间孝子鲍灿，坊阔9.54米，进深3.54米，高8.86米，建于明嘉靖年间，清乾隆年间重修。近楼的栏心板镂有精致的图案，梢间横坊各刻三攒斗栱，镂刻通明，下有高浮雕狮子滚球飘带纹饰的棚梁。四柱的磉墩，安放在较高的台基上。整座牌坊典雅厚重。慈孝里坊旌表宋代末年处士鲍宗岩、鲍寿孙父子，建于1501年，1777年重修。坊阔8.57米，进深2.53米，高9.60米。明间额枋较低，平板枋以上为仿木结构的一排斗栱支撑挑檐。明间二柱不通头。垫栱板朴质无华，加固了挑檐的基础，厚重相宜。鲍象贤尚书坊旌表兵部左侍郎鲍象贤，建成于明天启年间。四座清坊为鲍文龄妻节孝坊、鲍漱芳父子乐善好施坊、鲍文渊节孝坊、鲍逢昌孝子坊。四座坊均为冲天式，结构类似。大小枋额都不加纹饰，惟挑檐下的栱板，镂刻有花纹图案。月梁上的绦环与雀替也相应雕刻有精致的纹样。粗大的梁柱平琢浑磨，不事雕饰，另外在乐善好施坊、鲍文龄妻节孝坊间建有骢步亭，为四角攒尖式，清乾隆时期建，由邓石如书门额"骢步亭"。该石牌坊群以其独特组合、宏大规模及丰富的文化内涵成为中国建筑史上的一朵奇葩。

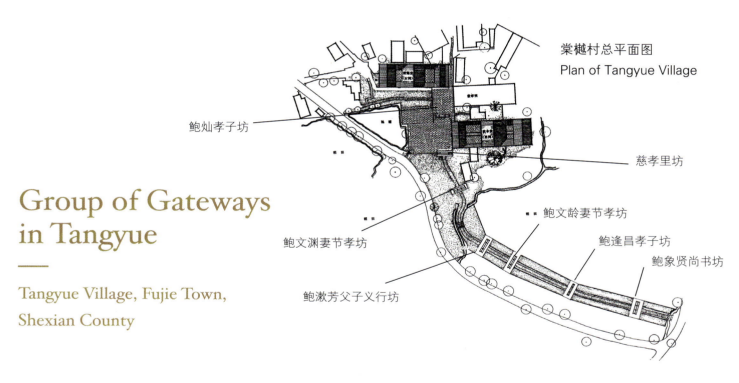

棠樾村总平面图
Plan of Tangyue Village

Group of Gateways in Tangyue

Tangyue Village, Fujie Town, Shexian County

The Group of Gateways are located on the main road east of Tangyue Village, Fujie Town, Shexian County. Of the seven gateways, three were built in the Ming Dynasty and four in the Qing Dynasty. These gateways can be divided, according to the contents and subjects, into four kinds of "Loyalty, Filial Piety, Chastity and Righteousness". The three Ming gateways are Bao Can Gateway, Cixiaoli (Lane of Love and Filial Piety) Gateway and Minister Bao Xiangxian Gateway. Bao Can Gateway was set up in the reign of Jiajing, in honour of Bao Can, a famous filial son, and rebuilt in the reign of Qianlong of Qing Dynasty. Cixiaoli Gateway was set up in 1501 and rebuilt in 1777, in honour of Bao Zongyan and his son Bao Shousun of late Song Dynasty. Bao Xiangxian Gateway was set up in the reign of Tianqi, in honour of Bao Xiangxian, a minister of the Board of War at the time. The four Qing gateways are: Bao Wenling's Wife Chastity Gateway, Bao Shufang and Son Charity Gateway, Bao Wenyuan Filial-piety Gateway and Bao Fengchang Filial-son Gateway. The four gateways are of the same style of four-column-soaring-to-the-sky and with few carvings and decorations. In addition, a pavilion was built in the reign of Qianlong, between the Chastity Gateway and the Charity Gateway. This series of gateways are marvelous spectacles of Chinese architecture for their unique composition, grand scale and rich cultural connotation.

棠樾牌坊群 the group of gateways

棠樾牌坊群 the group of gateways

四世一品坊 / 歙县雄村

四世一品坊位于歙县雄村。始建于明弘治和嘉靖年间，建于路一侧祠堂门前，四柱三间三楼，四柱冲天，梁、柱、额枋均茶园石雕凿而成，大小枋额都不加雕饰，明间花板镌刻着"四世一品"四个大字，花板下额枋刻有曹家父子及兄弟官至一品的功绩，明间正中有"圣恩"匾额，匾额之两侧盘以双龙祥云，整座牌坊气势宏大。

四世一品竖匾 The vertical inscribed board

Sishiyipin (Highest Court Officials for Four Generations) Gateway

Xiong Village, Shexian County

Sishiyipin Gateway is located in Xiong Village, Shexian County. The construction began in the reigns of Hongzhi and Jiajing of the Ming Dynasty. The gateway was built by the roadside, in front of the ancestral hall. It is of the style of four-column-soaring-to-the-sky and three bays, three towers. All the beams, columns and architraves are made of Chayuan stones, without carvings on them. On the panels there are inscriptions of the contributions of the Cao family.

四世一品坊

四世一品石坊 The stone gateway

四世一品匾额 The inscribed board

四面四柱石坊 / 歙县丰口村

四面四柱石坊位于歙县丰口村，建于明嘉靖年间，位于路边的一条溪水旁，是为旌表该村曾任"云南按察司佥事"的郑绮的功名而立，牌坊四柱四面，高11米，每边宽4米。这座牌坊的四柱摆在一个正方形的角上，每边都有楼、檐、梁、枋，构成了四个面，梁、柱为花岗石，枋、板为紫砂石。东面梁枋上无字，南面额枋上刻有"宪台"二字，垫板刻有"云南按察司佥事郑绮"一行小字。西面竖匾刻有"恩荣"二字，北面竖匾上刻有"勅赠"二字，额枋上刻有"廷尉"二字，垫板上刻有"大理寺左寺副郑廷宣"一行小字，脊檐下有华栱，竖匾左右有龙纹雕饰，檐枋下有的雀替有花卉和卷云雕饰。该坊在徽州牌坊中实属罕见。虽然每个面都像一座双柱牌坊，但整体看仍具有非凡的气势。

Four-side and Four-column Stone Gateway

Fengkou Village, Shexian County

This gateway is situated by a brook in Fengkou Village, Shexian County. It was set up in the reign of Jiajing of the Ming Dynasty, in honour of Zheng Qi for his contributions. It has four columns and four sides, 11 meters high 4 meters wide for each side. The four columns form a square corner, with each side of the same style, which is very rare in Huizhou gateways. Each side of it looks like a two-column gateway and the whole has a splendid view.

四面四柱石坊

四面四柱石坊全景 A full view of the stone gateway

昌溪木牌坊

歙县昌溪村

昌溪木牌坊位于歙县昌溪村，员公支祠前，建于清代中叶。该坊木质优良，榫卯结构，制作精密。四柱三间三楼，高8米，宽9.3米，四根木柱立在石础上，前后用八块夹杆石夹撑，上部有月梁、平板枋，枋上用斗栱承托歇山式盖顶，脊上砌透雕花砖，屋面盖小青瓦，高瓴垂脊，八角翘起，檐头饰以滴水。明间高出次间一层，错落有致，明间匾上书"员公支祠"四个大字。字匾左右的博风板红漆雕花，雕刻技艺精湛。是国内凤毛麟角的几个木牌坊之一。

昌溪木牌坊(背面) Wooden gateway (back)

Changxi Wooden Gateway

Changxi Village, Shexian County

Changxi Wooden Gateway is located in front of Yuangong Branch Ancestral Hall in Changxi Village, Shexian County. It was set up in the mid-Qing Dynasty, by using fine quality wood in a delicate construction of mortise and tenon. It is 8 meters high and 9.3 meters wide, with four columns, three bays and three towers. The four wood columns stand on the stone bases supported by eight pieces of stones. There is crescent beam, plate tiebeams, dougongs, brick-cared ridges covered with black tiles. An inscribed board of "Yuangong Branch Ancestral Hall" on the architrave is surrounded by delicate carvings. It is one of the few best wooden gateways in the country.

昌溪木牌坊 Wooden gateway

奕世尚书坊 / 绩溪县大坑口村

奕世尚书坊坐落在绩溪县的瀛洲乡大坑口村，建于明嘉靖四十一年。此坊位于一条溪水旁，有一石桥与其正对，为户部尚书胡富、兵部尚书胡宗宪共立。该坊系用花岗石和茶园石搭配雕凿而成，仿木结构，三间四柱五楼，高10米，宽9米。主体结构由四根柱子、四根定盘枋组成，气势雄伟，蔚为壮观。整体结构采用侧脚做法，向内微收。四根柱子由矩形石础承托，南北两向有抱鼓石夹护，并抹去棱角。四根定盘枋起线两道，再镌以莲瓣纹。梁柱接点处用花牙子雀替装饰。楼顶为歇山式，由斗栱支撑并挑梁。正脊两端鳌鱼对峙，明间正脊中置火焰珠。戗角腾空，戗兽架云。主楼正中装斜式"恩荣"匾，匾之两侧盘以双龙戏珠，其下方正面刻有"恩荣"、"奕世尚书"、"成化戊戌科进士户部尚书胡富"、"嘉靖戊戌科进士兵部尚书胡宗宪"、"大司徒"、"大司马"字样；背面刻有"恩荣"、"奕世宫保"、"太子少保胡富"、"太子少保胡宗宪"、"青宫少保"、"青宫太保"等文字。额枋部分的镂空圆雕尤其精美，大鹏展翅、仙鹤飞舞、狮子滚球、双龙戏珠等画面，栩栩如生，呼之欲出，额坊之花板上面镌刻着被立坊者和立坊者的官衔姓名。

Yishi Shangshu (Ministers) Gateway

Dakengkou Village, Yingzhou Town, Jixi County

This gateway is situated in Dakengkou Village, Yingzhou Town, Jixi County. By the side of a brook and facing a stone bridge, it was built by two ministers of the court—Hu Fu and Hu Zongxian, in the forty-first year of the reign of Jiajing in the Ming Dynasty. It was carved with granite and Chayuan stone in an imitation of wood-frame construction, four columns, three bays and five towers. The main structure consists of four columns and four pieces of Dingpanfang, and each of the columns has a rectangular stone base and is supported by two drums stones. The top of the tower is xieshan style, with Dougong supporting the eaves. There are inscribed boards and architraves with pierced carvings, from which we can see the names of those honoured and the sponsors.

奕世尚书坊全景 A full view of the gateway

胡文光刺史坊／黟县西递村

胡文光刺史坊位于黟县西递村前。建于1578年，清乾隆、咸丰年间曾修葺。是为"旌表"明嘉靖三十四年（1555年）科奉直大夫、胶州刺史胡文光而立，坊高12.3米，宽9.95米，四柱三间五楼仿木结构。梁、柱、额枋等均为质地坚实细腻的"黟县青"石料雕凿而成。全坊以四根60厘米见方抹角石柱为整体支柱，上雕菱花图案。柱下有长方形柱墩，各高1.6米，东西长2.8米，宽80厘米。明间两柱的靠背石前后都雕有一对倒卧石狮，形象逼真，威猛传神，高达2.5米，为支柱撑脚。明间额枋宽厚，枋心镂空透雕双狮滚球，还有三只仔狮盘球嬉耍，藻头为如意卷草纹和结带花等图案；两头次间的额枋上分别镂雕着凤凰、牡丹和松鹤图，雀替上也雕刻有云纹仙鹤；刻工纤秀细腻，线条刚劲流畅，造型逼真生动。中间横梁前后分别刻有"登嘉靖乙卯科奉直大夫朝列大夫胡文光"字样。二楼中间西面为"胶州刺史"、东面为"荆藩首相"斗大双钩楷字，书体遒劲。三楼中轴线上镌有"恩荣"二字，两旁衬以盘龙浮雕。二楼至四楼左右两侧和端点均流檐翘角，脊头吻兽雕为鳌鱼。左右次间石柱的靠背石为"雕日月卷象鼻格浆腿"式，形体卷曲流畅，次间都以雕花漏窗为饰，楼顶为悬山式，飞檐翘脊，檐下偷心栱板为一石凿成。外檐下斗栱两侧饰有44个圆形镂空花翅，四根石柱的东西两面共有12个穿榫，托着12块八仙、文臣武士人物雕塑，雕塑神态各异，精妙绝伦，是中国石牌坊中的精品之作。

胡文光坊匾牌 The inscribed board

胡文光坊局部石雕 Part of the stone carvings

Hu Wenguang Cishi (Provincial Governor) Gateway

Xidi Village, Yixian County

This gateway is located in front of Xidi Village, Yixian County. It was set up in 1578 and repaired in the reigns of Qianlong and Xianfeng in the Qing Dynasty, in honour of Hu Wenguang, provincial governor of Jiaozhou at the time of 1555. Carved with Yixian black stone, it is in an imitation of wood-frame construction with four columns, three bays and five towers. The gateway is framed of the four thick stone columns on the rectangular bases, with the main columns supported by the two huge lying stone-carved lions. On the architraves are carved vividly the design of lions, phoenixes, peonies and cranes. There are inscribed boards with carved characters of "Jiaozhou Cishi" and "Jinfan Shouxian", and various engravings. It is one of the treasures of stone gateways in China.

胡文光刺史坊

胡文光坊全貌　A full view of the gateway

方氏宗祠坊

歙县潜口村

方氏宗祠坊现移至歙县潜口民居博物馆，建于明嘉靖丁亥年间，四柱三间五楼单体仿木结构。通体为质地坚硬的"白麻石"雕刻而成，四根抹角石柱为整体支柱，柱下有长方形柱墩四个，柱的两侧各有一对巨大的抱鼓石支撑，各高1.9米，明间及两次间的额枋，均为镂空雕，刻有双狮戏球，凤戏牡丹，雕刻技艺精湛，柱梁间均用石栱承托，额枋间嵌以石雕漏花窗。中间横枋刻有"方氏宗祠"字样。值得称奇是二楼中间龙凤榜有一块雕刻"象形文字"的花板，花板左边雕有一鬼，右边雕了一个方形大斗，鬼和斗合在一起便是一个"魁"字，"魁"即是第一的意思，古时考取进士第一名称"魁甲"，考取举人第一名称"魁解"，由此可见，立坊者用心良苦。龙凤榜的背面雕刻有"月宫桂树"图案，寓意"蟾宫折桂"。各层楼左右两侧均流檐翘角，脊头吻兽雕为螯鱼。檐下斗栱两侧饰有36个圆形镂空花翅，四根石柱上还有4个穿榫，托着8个镂空花翅，整个石坊无论是花鸟鱼虫还是人物，均雕刻得细腻入微，令人叹为观止。

方氏宗祠坊局部石雕 Part of the stone carvings

Gateway of Fang Clan Ancestral Hall

Qiankou, Shexian County

This gateway has been removed into the Museum of Local Residence in Qiankou, Shexian County. Made of hard white stone, it was set up in the reign of Jiajing of the Ming Dynasty, in an imitation of wood-frame construction with four columns, three bays and five towers. The four columns on the bases form the main structure of the gateway, each supported by two huge drum-shaped bearing stones of 1.9 meters in height. On all the architraves are openwork carvings of lions and phoenixes. Between the architraves there are fretted windows and on the central horizontal architrave is an inscribed board with characters of "Ancestral Hall of Fang Clan". The whole gateway is full of delicate and exquisite carvings which are acclaimed as acme of perfection.

方氏宗祠坊

方氏宗祠坊正面　The gateway (front)

中国徽派建筑
HUIZHOU SCHOOL ARCHITECTURE OF CHINA

古塔

Ancient Pagodas

长庆塔

歙县西干山

长庆塔位于歙县城西练江南岸西干山，此处原有10座寺庙和长庆塔，其中长庆寺建于宋重和三年（1119年），如今其他寺庙皆毁，仅存该塔。该塔历代均有修葺。楼阁式，实心方形，高23.1米，底层平面每边5.28米，须弥座五层，束腰高66厘米，有间柱、角柱。塔身为砖砌，第一层较高，自下而上迭减。底层有木廊，石檐柱间宽4.33米。四面辟有券门，门内置石雕莲瓣佛座。第二层以上墙面中间均隐出窗券，各隅砌出半隐半露的方形角倚柱，墙面绘佛像彩色图案。每层檐口用砖叠涩挑出，间以五层斜角牙子。叠涩砖上为木构腰檐，覆以筒板瓦。飞檐翼角下，悬铁制风铃，长庆塔外形轻盈秀丽，是难得一见的古塔。

长庆塔全貌 A full view of Changqin Pagoda

Changqing Pagoda

the Xigan Hill in Shexian County

Changqing Pagoda is located on the Xigan Hill on the southern bank of the Xilian River in Shexian County. There used to be 10 temples and the Changqing Pagoda in this place and the pagoda was built in the third year of the Chonghe reign of the Song Dynasty (in 1119). Now all the temples have been damaged and this pagoda is the only relic. It was repaired almost in every dynasties since then. It is in a multi-storied solid square shape, 23.1 meters high for the whole and 5.28 meters wide for each side of the first floor. The xumizuo base is five-storied, and the suyao is 66 centimeters high with studdings and corner columns. The body of pagoda is made of bricks and the first layer is higher, decreasing progressively from the bottom to the top. On the first floor there is a wood corridor. It measures 4.33 meters between the stone peripheral columns. There are arch doors on the four sides and in the doors are stone-carved lotus blossoms. From the second floor up, the windows loom up in the middle of the walls. A square corner column is built in each corner. Colored designs of figures of Buddha were drawn on the walls. There are corbelling of bricks on every eave, covered with round tiles. Under the unturned eaves are hung iron wind bells. Changqing Pagoda is lightly and beautifully shaped; and it is a rare ancient pagoda.

新州石塔 / 歙县新州

新州石塔位于歙县城北郊新州,又名大圣菩萨宝塔。建于南宋建炎三年(1129年),用赭色麻石砦砌面成。重楼式,连顶部共五层,高4.6米,棱形挑檐,每层高度不同。塔为八面形体,第二层有香火炉窟,第三层左右两侧镌有斗大"佛"字,正面刻有南宋建炎建塔和明嘉靖年间重修的铭记,第四层八面均为如来神位字样,第五层发券内为如来佛像浮雕。该塔是乡人为祈求子嗣而自愿捐资建造的,简洁、质朴、没有任何花纹图案装饰,显得庄重、古朴。

新州石塔全貌 A full view of Xinzhou Stone Pagoda

Xinzhou Stone Pagoda

the northern suburb of Shexian Town

Xinzhou Stone Pagoda is located in Xinzhou in the northern suburb of Shexian Town, and it is also named Pagoda of Saint Buddha. It was constructed with the brown sienna stone boards in the third year of the Jianyan reign of the Southern Song Dynasty. It is multi-storeyed and altogether five storeys including the top. The pagoda is 4.6 meters high with every layer in different height. Having the prismatic projecting eaves, it is in the octagonal shape. On the second layer there is an incense burner. On the third layer a large character of "佛" is inscribed on the left and right walls and on the front wall is the inscription of the annals of the construction in the Jianyan reign of the Southern Song Dynasty and the reconstruction in the Jiajing reign of the Ming period. On the fourth layer, the words of "the tablet for Buddha" carved in all the eight sides. On the fifth layer, there are relieves of Buddha. This pagoda is contributed voluntarily by the villagers praying for son and heir. Without any decoration, it is simple and plain but solemn.

巽峰塔

休宁县下汶溪村

巽峰塔位于休宁县秀阳乡下汶溪村旁的玉几山上，建于明嘉靖年间，该塔为楼阁式砖塔，塔形八角七层，全高约35米，内有168级螺旋形梯道直通顶层，有塔心室，每层都有封闭的券门，菱牙砖叠涩出檐，平座以半砖挑出，塔顶有宝葫芦状塔刹，塔内有不少壁画，虽年代久远，线条仍清晰可辨。

巽峰塔全貌 A full view of Xunfeng Pagoda

Xunfeng Pagoda

Xiuyang Town

Xunfeng Pagoda is on the Yuji Hill near Xiawenxi Village in Xiuyang Town, Xiuning County. It was set up in the Jiajing reign of the Ming Dynasty. This is a pavilion-like brick pagoda of seven storeys in octagonal shape, 35 meters in height. There is a spiral staircase of 168 steps inside leading to the top. The inside chambers have arch doors on every floor. The eaves are rhombus bricks corbelling. The flat seat is projecting by half of a brick. The tee is calabash-like. The frescos in the pagoda are ancient but the lines are still clear and distinct.

丁峰塔

休宁县下汶溪村

丁峰塔位于休宁县下汶溪村旁的玉几山西，又名停风塔，建于明嘉靖年间（1544年），与巽峰塔遥相呼应，该塔为楼阁式砖塔，塔形六边五层，全高约30米，实心塔，每层都无门无窗，菱牙砖叠涩出檐，具有明显的辽代风格，虽年代久远，塔刹已毁，但其他部分保存完好。

丁峰塔全貌　A full view of Dingfeng Pagoda

Dingfeng Pagoda

Xiuning County

Dingfeng Pagoda is standing on the west of the Yuji Hill near Xiawenxi Village in Xiuning County. It is also named Tingfeng (Stopping the wind) Pagoda. Constructed in the Jiajing reign of the Ming Period (1544), it acts in cooperation with the Xunfeng Pagoda from the distance. This is a pavilion-like solid brick pagoda of five storeys in the hexagonal shape. It is 30 meters high without any door or window. The eaves are rhombus bricks corbelling. All these reflect the clear style of the Liao Period. Though it is of the remote past and the tee was destroyed, the rest is in good preservation.

富琅塔

休宁县万安镇

富琅塔位于休宁县万安镇富琅村。建于明万历年间（1594年），为楼阁式砖塔。八角七层，现存四层，外形完整，残高约17米。壁内有折上式楼梯，有塔心室，腰檐以菱牙砖叠涩出檐，平座以半砖挑出，每层券门上隐出额枋和普拍枋，额枋下隐出丁头栱，砖砌突伸的重檐及丁头栱，工艺精湛。塔砖长1尺，宽5寸，厚3寸，上有"万历癸巳宿"字样。

富琅塔全貌 A full view of Fulang Pagoda

Fulang Pagoda

the town of Wan'an,
Xiuning County

Fulang Pagoda is in Fulang Village in the town of Wan'an, Xiuning County. Constructed in the Wanli reign of the Ming Period (1594), it is a pavilion-like brick pagoda in octagonal shape. There used to be seven floors but only four are left in its integrity. The remains is 17 meters high. Inside there is a staircase and inner chambers. The middle eaves are the rhombus bricks corbelling. The flat seat is projecting by half of a brick. The architrave and pupaifang could be seen above the arch door on every floor. T-head dougong looms under the architrave. The T-head dougong and the multiple eaves laid by bricks are made in exquisite workmanship. The bricks are one chi long, five cun wide and three cun thick with the characters of "万历癸巳宿" on them.

古城塔

休宁县万安镇

古城塔位于休宁县万安镇古城岩。建于明嘉靖年间。为楼阁式砖塔，六角七层，底层以红麻石为基，基围21.3米，全高约29.6米，底边宽约1.4米，两个塔门。内部全空到顶，内空直径3.37米。底层原有副阶，二层以上设转角倚柱、腰檐、平座，二至五层的腰檐用斗栱挑出，斗栱出两跳，坐在普拍枋上，券门上隐出额枋，五层以上的腰檐用菱牙砖叠涩出挑。塔刹为生铁铸成，重2400公斤，向东南倾斜，1958年坠落。

古城塔全貌　A full view of Gucheng Pagoda

Gucheng Pagoda

—

the town of Wan'an,
Xiuning County

Gucheng Pagoda is on the Gucheng Hill in the town of Wan'an, Xiuning County. Set up in the Jiajing reign of the Ming period, it is the pavilion-like brick pagoda of hexagonal shape. It has seven floors and the first floor has the base of red stone board. The base measures 21.3 meters in perimeter. The total height of the pagoda is 29.6 meters, and each side is 1.4 meters wide in the first floor. It has two doors and is hollow inside, its inside diameter inside being 3.37 meters. There used to be side steps in the ground floor and corner columns, middle eaves and flat seat up from the second floor. The middle eaves from the second floor to the fifth floor are Dougong. The architrave could be seen above the arch door. The middle eaves above the fifth floor are the rhombus bricks corbelling. The tee is cast of pig iron and weighs 2400 kilograms. It leaned southeastward and fell down in 1958.

神皋塔

歙县岩寺镇

神皋塔位于歙县岩寺镇，始建于明代末年，清康熙年间及道光年间曾作过修葺。塔八面七层，高约66米，底径为8米，向上逐层内收，底层塔檐外挑15厘米，用菱牙砖叠涩出挑，由底向上出挑逐步加大，至第七层时塔檐挑出达30厘米，这样每层檐口的水都直接落到地面，可谓古塔建筑上的一绝，塔内有阶梯扶壁而上，每层都有佛龛及金匾，且有券门，可对外眺望。1914年塔顶被雷击倒塌，现存仅是珠墩以下的塔身。塔东有一凤山台，与塔同期建造。

神皋塔全貌 A full view of Shengao Pagoda

Shengao Pagoda

the town of Yansi, Shexian County

Shengao Pagoda is in the town of Yansi, Shexian County. It was first constructed in the twenty-third year of the Jiajing reign of the Ming period and was repaired in the Kuangxi and Daoguang reigns of the Qing Dynasty. In the octagonal shape, it is 66 meters high with seven floors. The diameter of the bottom is 8 meters and decreases by every floor. The eaves of the first floor are projecting out 15 centimeters in the rhombus bricks corbelling. The overhanging is growing from the bottom to the top. On the top floor, the eaves are overhanging by 30 centimeters, so the rain water from the eaves on each floor would fall down to the ground directly. This structure is unique in the pagoda construction. Inside there is a staircase and on each floor there is a Buddha shrine and a gilded plaque. Through the arch door, the outside view could be seen. In 1914, the peak of the pagoda collapsed by lightning strike, only the body of pagoda below the zhudun remained. To the east, there is the Fengshan Terrace which was constructed at the same time as the pagoda.

下尖塔

歙县潜口村

下尖塔位于潜口村南公路旁，旧志作"文峰塔"，俗称"潜口锥"，建于明嘉靖年间（1544年），由竹溪汪道植谨立。塔七层八角，底层直径10米，逐层缩小，外观如锥，第一层四面砌须弥座，墙上绘有佛像，第二层壁间嵌砖雕楣匾，内镌"翼峰"二大字，旁署"嘉靖二十三年甲辰岁，竹溪翁汪道植谨立"。其余五层为实心。现塔檐及顶部已毁。

下尖塔全貌 A full view of Xiajian Pagoda

Xiajian Pagoda

the south of Qiankou Village

Xiajian Pagoda is located by the road on the south of Qiankou Village. In the old annals, it is called *"Wenfeng Pagoda"* and also has a local name of *"Qiankou Wimble"*. It was contributed by Wang Daozhi from Zhuxi and constructed in the Jiajing reign of the Ming Dynasty (1544). In the octagonal shape, the pagoda has seven floors. The diameter is 10 meters at the bottom and it decreases by every floor. The appearance is like a wimble. The four sides of the first floor are laid with xumizuo bases and on the walls are the drawings of Buddha. On the second floor, there are lintels of brick carvings on the internal wall, on which two characters of *"Yifeng or wing peak"* are inscribed. Beside the characters is the inscription of the time of construction and name of the sponsor. The other five layers are solid. Now the eaves and the top are destroyed.

中国徽派建筑
HUIZHOU SCHOOL ARCHITECTURE OF CHINA

古桥

Ancient Bridges

洪桥 \ 歙县岩寺镇

洪桥北立面 North elevation of Hong Bridge

Hong Bridge

Yansi Town, Shexian County

Hong Bridge is located on the back street of Yansi Town, Shexian County. It was built over the Ying River in the Chenghua reign of the Ming period by the clansman Zheng Yanrong. At the end of the Yongzheng reign of the Qing period it was repaired with the donation from Zheng Weishu and Fang Dexian. The bridge is flat with three arches. There is a gallery of five bays. On the up side of the gallery there are flying chairs. The bridge is 17.4 meters long, 2.8 meters wide and 6.0 meters high. There is a chamber on each end. They are said to be the residence of the bridge keeper.

洪桥

洪桥坐落于歙县岩寺镇后街，建于明代成化年间，横跨于颍水之上，为族人郑彦荣所建，在清雍正末年郑为输和方德先捐资修葺过。该桥为三孔平桥，建廊屋五楹，临上游一面转置有飞来椅，桥长17.4米，桥面宽2.8米，高6.0米，桥两端各有小屋一间，传说为护桥者的居所。

洪桥南立面 South elevation of Hong Bridge

中国徽派建筑　古桥

洪桥内景　Internal view of Hong Bridge

洪桥

洪桥入口 Entryway of Hong Bridge

洪桥出口 Exit of Hong Bridge

彩虹桥

婺源县清华镇

彩虹桥位于婺源县北34公里的清华镇。桥横跨于古镇北侧头河上，桥初建于南宋，因袭唐诗"两水夹明镜，双桥落彩虹"而得名，后历代都曾加以修缮。桥为长廊式人行桥，全长140米，宽6.5米。桥基四墩呈半截船形，用青石叠砌而成，墩前端植有绿草花卉，墩尾是粉墙阁亭，亭中设有石桌石凳，盛暑供人纳凉歇晌、弈棋品茶。两桥墩之间跨径为15米，均以四根木梁横联，木梁上铺杉木板成桥面；木椽青瓦结顶，廊亭两侧有围栏和长凳供行旅凭眺憩息。桥上多楹联，如"胜地著华川，爱此间长桥卧波，五峰立极；治时兴古镇，尝当年文彭篆字，彦槐对诗"、"清景明时，彩画辉煌恢古镇；华装淡抹，虹桥掩映小西湖"等，桥中阁亭神龛内，供奉三个牌位，第一位是四处化缘筹资建桥的和尚胡永祥，第二位是后来捐资重修"彩虹桥"的胡永班，第三位是远古为治水三过家门而不入的大禹。另外据当地传说：桥落成之日，有彩虹悬于蓝天，天上地下，双景媲美，蔚为壮观。

Rainbow Bridge

Qinghua Town of Wuyuan County

The Rainbow Bridge is in Qinghua Town, 34 kilometers to the north of Wuyuan County. It was built in the Southern Song Dynasty over the Cetou River on the north of the ancient town, named after a poem of the Tang Dynasty. Repair was done in every dynasty afterwards. It is the footbridge with corridors, 140 meters long and 6.5 meters wide. It has four piers of a half-boat shape and laminated with gray stones. On the front of the piers, grass and flowers are planted and at the end there is a pavilion with whitewashed walls. In the pavilion are the stone tables and benches for people to enjoy the cool, to take a rest, to play chess and to sample tea in the hot summer. It measures 15 meters between two piers. Four wood beams span on each pair of piers and on the beams the deck is made of the fir plaques. The roof is made of wood rafters and gray tiles. Along the corridors there are railings and pews on both sides for passengers to rest and appreciate the scenery. There are many couplets on the bridge. There are three tablets In the shrine in the pavilion, there are three tablets respectively for Hu Yongxiang, a monk who begged alms for the construction of the bridge, Hu Yongban, who donated for the repair of the bridge, and for Yu the Great, who was too busy to go home—even passing by the door for water-control in ancient time. In addition, legend has it that when the bridge was completed, a rainbow hung over the sky. It was the spectacularity that the bridge and the rainbow echoed each other in harmony.

彩虹桥内神龛　Shrine in the bridge

彩虹桥全景　A full view of the Rainbow Bridge

中国徽派建筑　古桥

彩虹桥外景　External view of the Rainbow Bridge

彩虹桥

中国徽派建筑
HUIZHOU SCHOOL ARCHITECTURE OF CHINA

古
亭

Ancient
Pavilions

文昌阁

歙县雄村

文昌阁位于歙县雄村竹山书院内，建于清乾隆年间，为曹翰屏所建。平面呈八角形，俗称"八角亭"，高二层，筑于高台之上，首、二层均为八根檐柱，二层有外廊，八面门扇均为通花门扇，八条垂脊末端翘起，飞檐筒瓦，戗角处横铺蝴蝶瓦，戗角下悬金雀铃，阁楼攒尖顶。葫芦形锡顶银光熠熠。南面楼檐下悬"贯日凌云"四字金匾。楼内藻井、梁枋彩绘灿然。原先供有文昌菩萨，以显示家族的文风昌盛。

Wenchang Pavilion
– The Pavilion of Flourishing Culture

Xiongcun Village, Shexian County

Wenchang Pavilion is located in the Academy of Bamboo Hill in Xiongcun Village, Shexian County. It was constructed by Cao Hanping in the Qianlong reign of the Qing Dynasty. The plan is in octagonal shape, so the local name of it is "Octagonal Pavilion". Based on the high platform, it has two storeys. There are eight peripheral columns on each storey and on the second floor there is a side corridor. All the eight door-leaves are decorative openwork. The ends of the eight drooping ridges turn up, covered with upturned eaves and cylindrical tiles. The hip is paved with Chinese convex and concave tiles. Below the hip are hung the canary-like bells. The top is the polygonal roof. The calabash-like tee is blocked with tin and shining. Below the southern eaves the gilded plaque is hung with four characters of "*Guan Ri Ling Yun*". The colored drawings on the caisson and girders are still bright and splendid. The God of Literature used to be worshiped inside to show that the clan had a tradition of scholars.

文昌阁立面 Facade of Wenchang Pavilion

绿绕亭

歙县西溪南村

绿绕亭位于徽州区西溪南村老屋阁东南墙脚下池塘畔。建于1328年，1456年重修。亭平面呈正方形，通面阔4米，进深4.36米，高5.9米，跨街而建。亭结构与雕饰风格类似老屋阁，为抬梁式屋架，梁架上有"明景泰七年重建"字样。惟月梁上绘有包袱锦彩绘图案，典雅工丽，有元代彩绘遗韵。亭临池一侧置"飞来椅"。在亭中近可观繁茂场圃，远可眺绿茵田畴。明代著名书画家、大才子祝枝山曾作"东畴绿绕"一诗赞咏绿绕亭的绮丽风光：

庞公宅畔甫田多，畎亩春深水气和。

五两细风摇翠练，一犁甘雨展青罗。

鱼鳞强伏轻围径，燕尾逶迤不作波。

Lürao Pavilion
– The Pavilion of Greenery Winding

Xixi Village, Huizhou District

Lürao Pavilion is standing beside the pond near the southeast wall of the Chamber of Old-House in the South Xixi Village, Huizhou District. It was built in 1328 and rebuilt in 1456. The pavilion is square in plan, 4 meters in breadth, 4.36 in depth and 5.9 meters in height, standing across the street. The structure and the style of decorations are similar to the Chamber of Old-House; the frame is post and lintel construction and on the beam structure is written the words of " Reconstructed in the Seventh Year of the Jingtai Reign in the Ming Dynasty". On the crescent beam are coloured paintings of Yuan style. At the side by the pond are flying chairs. People can enjoy the gardens nearby and the green fields in distance. Zhu Zhishan, a famous painter and scholar of the Ming Dynasty, once wrote a poem in praise of the pavilion.

绿绕亭全貌 A full view of Lürao Pavilion

绿绕亭梁架 Beam structure of Lürao Pavilion

沙堤亭

歙县唐模村

沙堤亭位于歙县唐模村村口,建于清康熙年间,跨路而建,亭为三层,平面为方形,边长为6.1米,底层四边十二檐柱均为石柱,并置有石凳,四边各有一门,进村的道路穿过该亭,路人可在亭内休息和避雨,二层为虚阁,四面设有栏杆,屋面三重檐,歇山顶,外置挂落,飞檐翘角,悬挂风铃,饰龙吻。垂脊上饰有天狗、鸡禽等兽,亭东西门额上题有"沙堤"和"云路"的匾,一层的檐下还有一块"风雨亭"的匾额。远看沙堤亭,整个建筑清秀美观,似一婷婷玉女站在树下。

沙堤亭"风雨亭"匾额 Inscribed board of "Storm Pavilion"

沙堤亭全貌 A full view of Shadi Pavilion

Shadi Pavilion
– The Pavilion of Sand-Bank

Shexian County

The Pavilion of Sand-Bank is located at the entrance of Tangmo Village in Shexian County. Constructed in the Kangxi reign of the Qing Dynasty, it stands across the road with three floors. The plan is square and each side measures 6.1 meters. Twelve peripheral columns on the four sides of the first floor are made of stone. There are stone benches inside and a door on each side. The road to the village passes across the pavilion. Passengers can have a rest and it is a shelter from rain. The second floor is virtual with railings around. The top is triple-eave xieshan roof with hanging fascias outside. The eaves are upturned with decorations of dragon heads and wind bells are hung out. The drooping ridges are furnished with the heavenly hounds, fowls and birds. The plaques of "Sand Bank" and "Clouds Path" are hung on the east and west head jambs. Under the eaves of the first floor there is an inscribed board of "Wind and Rain Pavilion". The pavilion looks comely and beautiful from the distance, like a graceful lady under the tree.

沙堤亭全貌 A full view of Shadi Pavilion

善化亭 \ 歙县许村镇

善化亭位于歙县许村镇"现移至潜口民居博物馆",建于明嘉靖辛亥年(1551年),系里人许岩保捐资所建,意在行善,故名"善化亭"。该亭呈四方形,亭内花岗石铺地,两侧置有花岗石石凳,供行人歇息,亭顶为歇山式,正脊和翘角饰有龙吻,斗栱的补间为"一斗三升",柱头为"斗口跳",脊檩上依然可见当时的题字"嘉靖辛亥春许村许岩保偕室宋氏喜彼杨充岭石路雨亭,以便往来福佑攸归者"。两道三架梁上书有"阳春有脚九重天上行来,阴德无根方寸地中种出"的对联,充分体现了当时人们对积德行善的称颂。

Shanhua Pavilion
– The Pavilion of Mercy

Shexian County

Shanhua Pavilion was located in Xucun Town in Shexian County, but now it is moved to Qiankou Museum of Folk Residence. Constructed in 1551, in the reign of Jiajing of the Ming Dynasty, it was donated by the villager Xu Yanbao to display his mercy, hence the name. This pavilion is square and paved with granites. The granite benches are placed on two sides for passengers to rest. The top is the xieshan roof. There are dragon-head ornaments on the main ridges and upturned angles. On the ridge purlins some inscriptions of the pavilion construction still can be seen. A pair of couplets are on the beam in praise of people's mercy behaviours.

善化亭梁架 Beam structure of Shanhua Pavilion

善化亭全貌 A full view of Shanhua Pavilion

魁星阁

绩溪县旺川石家村

魁星阁位于绩溪县旺川乡石家村。建于清康熙末年（1715年）。阁楼基高0.7米，阁高2.5米，楼顶采用七分水法，四面落檐，落地檐高1.7尺，楼台四角离地19尺，每方用椽50根，阁下正面上方，原有一块横匾，上题"魁星阁"三字。匾的上方还有一尊魁星像。阁左侧有一长6米，宽、高3.3米的土石平台，传说象征石家村始祖，北宋开国元勋石守信的帅印。平台中间栽一枫树，犹如"印柄"。该阁布局独具匠心，据说还蕴含"反清复明"的深意。

Kuixing Pavilion
– The Pavilion of God of Literature

Wangchuan Town of Jixi County

Kuixing Pavilion is located in Shijia Village in Wangchuan Town of Jixi County. It was built at the end of the Kangxi reign of the Qing period (1715). The base is 0.7 meter high and the pavilion is 2.5 meters high. The four corners of the top floor measure 19 chi high from the ground, and on each side there are purlins. In the front of the pavilion there used to be an inscribed board with the characters of "Kui Xing Ge", and a figure of God of literature above the board. To the left of the pavilion there is a platform of earth and stone, 6 meters in length and 3.3 meters in breadth and height. Legend has it that this is the seal of Shi Shouxin, a founder of the Northern Song Dynasty and earliest ancestor of Shijia Village. A maple tree, just like the handle of the seal, is in the middle of the platform. This structure shows the originality and it is said to have the meaning of "Recovering the Ming against the Qing".

魁星阁

魁星阁立面　Façade of Kuixing Pavilion

魁星阁全貌　A full view of Kuixing Pavilion

中国徽派建筑
HUIZHOU SCHOOL ARCHITECTURE OF CHINA

古书院

Ancient Academies

竹山书院

歙县雄村

竹山书院位于歙县城南6公里的雄村。系清乾隆年间曹翰屏所建，砖木结构，建筑面积1130平方米。门外有空场，场边为桃花坝，坝下即浙江。主体为一厅式建筑，二进三楹，厅堂进门是前廊，隔天井为三开间后堂。右廊有一侧门，通向内院。这里既有教室也有先生的书斋和住宿、活动用房。中间辟有小院、花圃。廊道尽头，有庭园，名"清旷轩"，系一小型古典园林。当时曹氏族约："子弟中举者可在庭中植桂一株"，故又名桂花厅，厅内仍遗有桂花树数十棵。还存有乾隆时期大诗人曹学诗的"清旷赋"屏障，书法家郑莱的"所得乃清旷"小篆匾额，及摹刻颜书"山中天"石刻。

竹山书院"山中天"匾额
Inscribed board of "Shan Zhong Tian" (Sky in the Mountains)

竹山书院"所得乃清旷"匾额 Inscribed board

Zhushan Academy
– The Academy of Bamboo Hill

Xiongcun Village, Shexian County

The Academy of Bamboo Hill is located in Xiongcun Village, 6 kilometers south to Shexian County. It was constructed by Cao Hanping in the Qianlong reign of the Qing Dynasty. It is the brick-and-timber construction and covers an area of 1130 square meters. There is a yard in front of the Academy and the Peach-blossom Dam. Below the dam is the Zhejiang River. The main body of the academy is the one-hall construction. It has two sections and three bays. Next to the lobby is the front corridor. The three-bay rear hall is separated by the skywell. There is a side door on the right corridor, leading to the inner yard. There are classrooms, study of the teacher and rooms to live and play. In the middle are the small yard and garden. At the end of the corridor is a courtyard, a mini-type classical one named "Qingkuang Veranda". According to the agreement of that time — "Any family member who passes the provincial examination can plant a laurel tree in the garden", the place is also named Laurel Hall, with totally tens of laurels still left. Inside it there is a screen on which is inscribed a poem written by Cao Xueshi, a famous poet in the Qianlong reign, an inscribed board written by a well-known calligrapher Zheng Lai, and a stone inscription of "Sky in the Mountains" in Yan Zhenqing style.

竹山书院全貌 A full view of Zhushan Academy

竹山书院清旷轩 Qingkuang Veranda in the academy

古紫阳书院 \ 歙县城内

古紫阳书院位于歙县县城华屏山南坡。始建于南宋嘉定十五年（1222年），时名文公祠，后残破不堪，清乾隆五十五年（1790年），由邑人曹文埴倡议复建。一批重视儒学的徽商集资于旧址重建书院，建筑物近1800平方米，中轴线上前为朱子殿，中为尊道堂，后为韦斋祠。左为据德舍、志道舍，右为依仁舍、游艺舍，凡是为教学用房，用甬道间隔。朱子殿前坡下为考棚；左为院门，外有"古紫阳书院"石坊；右有文公井，为郡城名泉之一,在朱子殿内，存有清康熙三十二年（1693年）皇帝御书的"学达信天"匾和乾隆九年（1744年）御书的"道脉薪传"匾和"百世经师"匾。以及乾隆五十五年著名学者程瑶田书写的"古紫阳书院规条"石刻。东南甬道上有一座由曹文埴题额的"古紫阳书院"石门坊，这座书院是为纪念理学大师朱熹而建的，南宋以后理学盛行，得到历代皇帝的推崇，"紫阳书院"院名乃宋理宗所赐。

The Ancient Ziyang Academy

the town of Shexian County

The Ancient Ziyang Academy is on the southern hillside of the Huaping Mountain in the town of Shexian County. It was first built in the fifteenth year of the Jiading reign of the Southern Song Dynasty (1222), and named "Ancestral Hall of Wengong" at that time. Later it was dilapidated. In the fifty-fifth year of the Qianlong reign of the Qing period (1790), it was rebuilt at the proposal of Cao Wenzhi. A group of businessmen of Huizhou who valued Confucianism donated to rebuild the Academy at the site covering an area of 1800 square meters. Along its central axis there are first Zhu Xi Temple, then Zundao Hall or The Hall of Revering Teachings, and the last Fengzhai Hall. On the left side are Jude Room, Zhidao Room, and on the right are Yiren Room and Youyi Room. There are corridors between the classrooms. Down the slope in front of Zhu Xi Temple are the examination chambers with the gate on the left. A stone arch stands outside, with the inscription of "The Ancient Ziyang Academy"; on the right is Wengong Well, which is the famous spring in the county. Inside the Zhu Xi Temple, there are the inscribed board of "Xue Da Xin Tian" written by the Kangxi Emperor himself in the thirty-second year of the reign (1693), and another two of "Dao Mai Xin Chuan" and of "Bai Shi Jing Shi" in the ninth year of the Qianlong reign in the Qing Dynasty (1744), together with the stone inscription of "Regulations of the Ancient Ziyang Academy" written by the notable scholar Cheng Yaotian in the fifty-fifth year of the Qianlong reign. On the southeast corridor is a stone gate arch of "The Ancient Ziyang Academy" inscribed by Cao Wenzhi. This academy was built in commemoration of the great Confucian scholar Zhu Xi. After the Southern Song Dynasty the Confucian school of idealist philosophy was in popular and was canonized by all the emperors afterwards. The name of "Ziyang Academy" is bestowed by Emperor Li Zong of the Song Dynasty.

古紫阳书院石坊 Stone gateway in Ziyang Academy

古紫阳书院入口 Entrance

古紫阳书院学舍 Dormitories of Ziyang Academy

古紫阳书院学舍 Dormitories of Ziyang Academy

文庙书院 \ 绩溪县华阳镇

文庙书院位于绩溪县华阳镇，据清嘉庆《绩溪县志》载，文庙始建于宋。元至元十三年（1276年）毁于战乱，至大元年（1308年）重建。明正德七年（1512年）对文庙进行大修，使其在原有基础上再进一步扩增，完善了文庙建筑的平面布局及单体建筑。明嘉靖三十九年（1560年），邑人少保胡宗宪捐资对文庙再次整修扩建。清乾隆四十二年（1777年）又重建文庙，历时八年告竣，沿南北中轴线东西对称布置建筑，依次有大成殿、露台、东西两庑、斋房、戟门、斋明所、宰牲所、泮池、泮桥、泮宫坊、棂星门。自宋至清，文庙先后经过了二十次修葺扩建。现除大成殿、东西两庑、泮池、泮桥外，其他建筑已不存，但基址仍在，其总体格局仍可一目了然。

大成殿建于清乾隆年间，坐北朝南，面阔七间，进深三间，外设廊，建筑面积260平方米。重檐歇山顶，中设腰檐，四路花砖正脊，两端鸱吻，脊中部置宝瓶。梁架上层为草栿，天花下为明栿，用材硕大，制作精良。精美的斗栱将两层屋檐托起，使得整个屋面轻巧明快，线条柔和流畅。殿内悬挂清康熙二十三年（1684年）的"万世师表"匾额，其下设置祭案。殿的东西及后檐墙体均漆以朱色，与青瓦屋顶形成和谐的色彩对比。整座大殿气势雄伟，蔚为壮观。殿前是一近100平方米的青石露台，其东西及正面围护石栏杆，栏板上镌以山水、花鸟画面。东西两庑面阔七间，进深三间，外为廊，建筑面积388平方米，硬山式屋顶，梁架用材较大成殿为小，铺设天花，未置斗栱，墙体亦漆以朱色，与主体色调协调一致。

Wenmiao Academy
– The Academy at the Temple of Confucius

Huayang Town, Jixi County

Wenmiao Academy is located in Huayang Town, Jixi County. According to Annals of Jixi County in the Jiaqing reign of the Qing Dynasty, it was first constructed in the Song period. In the thirteenth year of the Zhiyuan reign (1276), it was damaged in the war and then rebuilt in 1308. In the seventh year of the Zhengde reign of the Ming Dynasty (1512), it was completely repaired and expanded on the original basis. Its overall plan arrangement and single constructions are perfected. In the thirty-ninth year of the Jiajing reign of the Ming period (1560), it was refitted and expanded with the donation from Ha Zongxian. In the forty-second year of the Qianlong reign (1777), it was constructed again and it took eight years to complete the construction. Along the central axis from north to south, the arrangement of the buildings is symmetrical—Dacheng Hall, the Platform, the Eastern and Western Shrines, the Halberd Gate, the Abstinence Rooms, the Slaughterhouse, the Pool of Pan, the Dissolving Water Bridge, the Arch, and the Lingxing Gate. From the Song period to the Qing period, the Temple of Confucius was repaired and enlarged twenty times. Facing south, Dacheng Hall was built in the Qianlong reign of the Qing period. It is seven-bay in width and three-bay in depth with corridors outside on an area of 260 square meters. The top is the multiple-eave xieshan roof. The four main ridges are covered with tiles, with chiwens (owl-head ridge ornament) at both ends and treasured bottles in the middle. The grass roof beam is on the top of the beam frame and under the ceiling is the exposed beam, both of which are huge and delicately made. The exquisite dougongs support the two layers of eaves to make the room light and bright, and the lines soft and smooth. The plaque of "The Teacher for All Ages" was hung in the twenty-third year of the Kangxi reign of the Qing Dynasty and below it was the sacrificial table. The east, west and rear walls of the Hall were painted in red in harmonical contrast with the roof of black tiles. The vigor of the Hall is imposing and magnificent.

文庙书院

文庙书院外景 Exterior view of the acedemy

文庙书院连廊 Corridors

文庙书院大成殿脊梁 Ridged beams of Dacheng Hall

考棚

绩溪县华阳镇

考棚坐落在绩溪县华阳镇绩溪中学校园内。根据文昌殿石碑记载推断，明伦堂考棚约始建于清初。原规模较大，有头厅、候考厅、东、西考棚、明伦堂、魁星阁、藏经楼等建筑。现仅存头厅、庭院、候考厅。该建筑坐北朝南、砖木结构、硬山屋顶，山墙砌筑封火墙，撑栱承挑出檐。头厅面阔五间，进深一间，设有八字形大门，其后是庭院。候考厅面阔五间，进深二间。总占地面积374.5平方米。八字门的墙体上绘制彩画，木结构的驼峰、雀替、撑栱等均有雕刻图案。在候考厅西山墙嵌有石碑一块，"为重修明伦堂考棚各工程阖邑捐输姓名详数刊列于后"落款是"光绪八年（1882年）三月"。1990年全面维修，并移山水石雕栏板六块安装在候考厅前檐。

The Examination Chambers

Huayang Town, Jixi County

The Examination Chambers are on the campus of Jixi Middle School in Huayang Town, Jixi County. Deduced from the records on the stone stele in Wenchang Hall, the Examination Chambers of Minglun Hall was built at the beginning of the Qing period. The original was in large scale with a front hall, a waiting hall, the east and west examination chambers, Minglun Hall, the Pavilion of God of Literature and the Tripitaka Pavilion. But now only the front hall, the yard and the waiting hall are left. Facing south, it is the brick-and-timber construction with yingshan roof or the Chinese gabled roof. The walls are fire-sealing gables and the eaves are projecting. The front hall is five-bay in breadth and one-bay in depth with a splay door. Behind it is the yard. The waiting hall is five-bay in width and two-bay in depth, covering an area of 374.5 square meters. On the splay walls are the colored drawings and there are inscriptions on the wood hump-shaped pad blocks, queti and brackets. On the west wall of the waiting hall is embedded a stone stele, with the name list of the sponsors and the time of construction (March, 1882). In 1990, it was completely repaired and six railing panels of stone inscriptions of scenery were moved under the front eaves of the waiting hall.

考棚内全景　The internal full view of the chambers

考棚八字门入口　8-shaped entryway

南湖书院 / 歙县宏村

南湖书院坐落于黟县宏村南湖北畔。书院始建于清代嘉庆十九年（1814年），占地6000余平方米。"南湖书院"，又名"以文家塾"，分左中右三大间，左侧悬挂"南湖书院"匾额，首进为木栅门厅；二进为书院正厅"志道堂"，系教书授业之处；三进为"文昌阁"，供奉的是孔子和朱熹的牌位。另外还有启蒙阁、会文阁、望湖楼和祗园四部分组成。"会文阁"，是宗族中文人墨客以文会友处，另有庭院、操场，占地达1公顷。

南湖书院外临一湖碧水，庭院内置花园假山，操场上有株百年龙柏。书院原有金色匾额"以文家塾"四字，为清朝翰林院侍讲、著名书法家梁同书年93岁时所书。西侧有一卷棚式屋顶的"望湖楼"，上悬匾额"湖光山色"四字，系清代歙县人、时任黟邑知县罗廷孚题，今仍完好。登楼极目四野，湖光山色尽收眼底。

Nanhu Academy
－The Academy by the South Lake

Hongcun Village, Shexian County

Nanhu Academy is on the north bank of the South Lake in Hongcun Village, Shexian County. The Academy was built in the nineteenth year of the Jiaqing reign of the Qing Dynasty (1814), covering an area of 6000 square meters. It is also named "Yiwen Family School". It has three large halls—the left, the middle, and the right. On the left is hung the board of "Nanhu Academy" and the first section is the lobby with wood fence; the second section is the main hall of the Academy—"Zhidao Hall" where lessons are taught; the third section is the Wenchang Hall where the tablets of Confucius and Zhu Xi are worshiped. In addition, there are four parts—Qimeng Hall (the Chamber of Enlightenment), Huiwen Hall (the Chamber of the Literary Circle), the Lake-Viewing Tower and the Garden of Respect. The Chamber of the Literary Circle was the place where the scholars of the clan met friends on literary. It covers an area of one hectare together with the garden and the playground.

Outside the Academy there is a lake. Inside the garden there is rockery and on the playground there is a dragon cypress tree of one hundred years. There used to be a gilded plaque of "Yiwen Family School", written by Liang Tongshu, a famous calligrapher when he was ninety-three years old. On the west side there is the Lake-Viewing Tower with juanpeng roof. The plaque of "Natural Beauty of Lakes and Mountains" hung there was written by Luo Yinfu from Shexian County, a magistrate of Yixian County in the Qing period. The tower is still in good condition. Mounting the tower, you can have a panoramic view of the landscape of the lakes and mountains.

南湖书院全貌 A full view of the academy

南湖书院"文昌阁" Wenchang Hall in the academy

南湖书院大门 Entrance

南湖书院"志道堂"　Zhidao Hall in the academy

中国徽派建筑
HUIZHOU SCHOOL ARCHITECTURE OF CHINA

古民居

Ancient
Folk
Residence

许国相府

歙县县城

许国相府位于歙县县城向阳路15号。建于明中叶，为许国府邸的部分遗构。坐东朝西，凹字形平面，一进二层，面阔13.94米，进深8.65米。大门开在左侧，两旁为廊房，中央为天井。山面穿斗式梁架，月梁雕刻华丽，双步梁端雕饰的构图下面用两朵小云承托上部的大云，有的大云前后两端作尖状。三架梁上立有脊瓜柱承托脊檩，两侧置雕花叉手，整个梁头很像一条飘带。不用平盘斗时，脊瓜柱下端咬杀成鹰嘴形。1987年，曾按原样进行维修。

许国相府后天井 The back courtyard

许国相府"对弈"石雕 The stone carving of chess game

Residence of Premier Xu Guo

Shexian County

Residence of Premier Xu Guo is located at No. 15 of the Xiangyang Road in the town of Shexian County. Constructed in the mid-Ming Dynasty, it is a part of the remains of the Residence of Xu Guo. Facing west, with a plan of concave, it has one section and two storeys, 13.94 meters in breadth and 8.65 meters in depth. The door is on the left and the two sides are the corridors with a courtyard in the middle. It has the through-jointed frame. The crescent beam is decorated with splendid inscriptions. There is a king post on the three-purlin beam to support the ridge purlin, and on both sides are the engraved inverted V braces. The whole top is like a ribbon and the lower part of the king post is of the shape of an eagle's hawk when no pingpandou is used. In 1987, the residence was rebuilt according to the original shape.

许国相府天井 The courtyard

许国相府后堂 The back hall

汪宅

歙县斗山街

汪宅位于歙县斗山街，建于清末。两进三开间堂屋，前堂三开间，左、右厢房和窗栏木雕精致、典雅，雕工细腻，笔法含蓄，有着深厚的意韵。在我所见众多的精美窗雕中，是最具有徽派木雕艺术特色的一家。如窗棂雕刻的"八仙图"，不见八位仙人人物形象，但雕刻铁拐李的葫芦，汉钟离的阴阳宝扇，张果老的渔鼓，何仙姑的荷花，蓝采和的花篮，吕洞宾的宝剑，韩湘子的竹笛和曹国舅的檀板。以物喻人，并衬之以琴棋书画，在这幅图上称之为"暗八仙"。下方的"五福拜寿"，更使画面充满吉祥喜气。我们可见对面的门雕上，有着活泼的九只松鼠，这就是"九松图"，"松鼠"的"松"与"子孙"的"孙"谐音，"九松图"实为"九孙图"，表示对多世同堂，子孙满室的祝福。此图取材为汉代贤士张公翼，他一生积德，高寿多孙。这些木雕中，还有十二个月花，四季果品，从这些木雕隐示，暗喻的图案中，可见徽商有着很高的艺术修养，处世不俗，不愧为一代儒商。

Residence of the Wangs

Doushan Street in Shexian County

汪宅正堂 The main hall

Constructed at the end of the Qing Dynasty, Residence of the Wangs is on Doushan Street in Shexian County. There are two sections of three-bay central rooms. The front section has three rooms. The woodcarvings on the window frames in the left and right wing rooms are exquisite and elegant. They are minute in craftsmanship and reserved in strokes with a profound charm. This is the most typical one in the characteristics of woodcarvings of Huizhou School. For instance, in the "Design of Eight Immortals" carved in the window lattice, the figures of the Eight Immortals could not be seen but their instruments—the gourd for Li Tieguai, the fan for Zhongli Han, the bamboo percussion instrument for Zhang Guolao, the lotus for He Xiangu, the flower basket for Lan Caihe, the sword for Lü Dongbin, the bamboo pipe for Han Xiangzi and the hardwood clappers for Cao Guojiu. All these instruments stand for their owners, together with the lute, chess, calligraphy and painting; so it is called "The Hidden Eight Immortals". Below them is the carving of "Five Bats" adding happiness to the whole, which means the blessing of long life. In the carvings on the opposite door, there are nine lively squirrels—Design of Nine Squirrels. The sound of "squirrel" in Chinese is similar to that of "grandson", so actually it implies "Design of Nine Grandsons" which means the blessings of many generations under one roof and many children and grandchildren. This design is based on the story of Zhang Gongyi, the virtuous and learned man in the Han Dynasty. He followed a virtuous path throughout his long life and had lots of descendents. Among these woodcarvings are flowers of twelve months and fruits of four seasons. From the designs of these carvings, it is obvious that businessmen in Huizhou District were artistically accomplished and not hackneyed in their ways of life. They proved themselves to be learned and refined businessmen of no equal of their time.

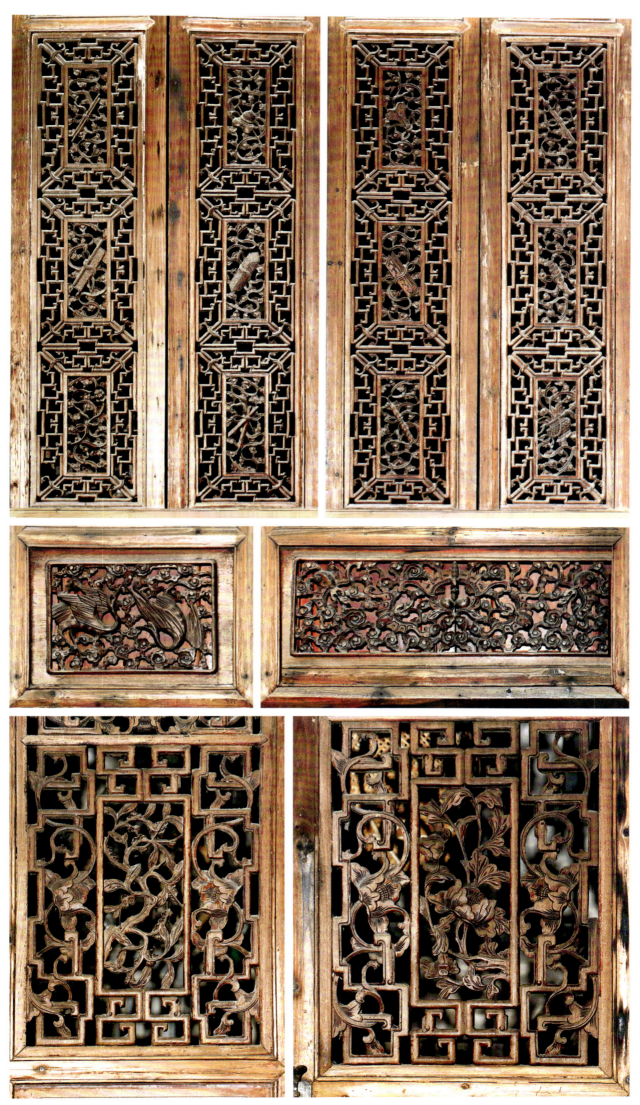

汪宅窗板 Window boards

汪宅入口 The entryway

汪宅

许氏大院

歙县斗山街

许氏大院位于歙县斗山街,建于清初。为二进一开间,入口为一窄小的棋门,进门后空间豁然开朗,正堂高五米有余,顶棚和梁架均有精美的彩绘,是典型的徽派古代私塾建筑结构,正厅的两侧柱子上挂着落地的匾额,香案上方高悬孔子的画像,正厅主要是用来讲学和祭孔的场所,楼上是学子读书的地方,穿过左侧的镜瓶门是先生的寝室,整个院落布局讲究,院中栽有多棵桂花树,采用了许多苏州园林的手法,别有一番儒雅风味。

The Compound of the Xus

Doushan Street, Shexian County

Built at the beginning of the Qing period, the Compound of the Xus is at Doushan Street. It has two sections of one-bay houses and the entrance is a narrow archway. Then the space becomes open and clear suddenly. The main hall is over five meters high. The plafond and the beam with beautiful colored drawings on them are typical structures for old private schools in Huizhou. On the side columns in the main hall are hung the down-to-floor couplets and over the incense burner table is hung the portrait of Confucius. The main hall was the place to give lectures and hold memorial ceremonies for Confucius. The second floor was the place for students to study in. Through the vase-shaped door on the left is the bedroom for the teacher. By adopting lots of techniques of Suzhou garden art, the arrangement of the compound is dainty and has an entirely learned and refined flavor.

许氏大院正堂 The main hall

黄宾虹故居

歙县潭渡村

黄宾虹故居位于歙县潭渡村。建于清中晚期。1876年黄宾虹从金华回歙县应童子试，在故里潭渡村生活约30年。故居正屋为三开间，共二层楼，前有庑廊和小天井，庑廊一边设有"美人靠"，坐此可观赏院内花草，堂匾和门额上为黄宾虹自题的手书"宾虹草堂"和"虹庐"。左廊题"冰上飞鸿馆"。屋前为小院。出左院门，为"玉森斋"，是一座三开间平房。前院有假山石一块，名"石芝"。黄宾虹常在画上题"写于石芝室"或"石芝阁"，即指此处。

黄宾虹故居庭院 The courtyard

Former Residence of Huang Binhong

Tandu Village, Shexian County

Constructed in the middle-late Qing Dynasty, the Former Residence of Huang Binhong is in Tandu Village, Shexian County. In 1876, Huang Binhong returned to Shexian County from Jinhua to take the preliminary examination and lived in his hometown—Tandu Village for about thirty years. The main house has two stories of three bays. In front of it are the corridors and a small skywell. There are beauty chairs on one side of the corridors, where the plants in the garden can be viewed. On the boards of the hall and the head jamb are the inscriptions of "Cottage of Binhong" and "Hong's Hut" of his own handwriting. There is a yard in front of the house. Outside the left door is "Yusen Zhai", which is a three-bay bungalow. There is a rockwork named "Stone Glossy Ganoderma". When Huang Binhong wrote on the paintings "Written in the Room of Stone Glossy Ganoderma" or "Chamber of Stone Glossy Ganoderma", he referred to this place.

黄宾虹故居入口 Entrance

黄宾虹故居"宾虹草堂"内景 The internal view of "Cottage of Binhong"

黄宾虹故居

Old House

老屋阁／歙县西溪南村

Xixinan Village of Shexian County and originally the residence of Wu Xizhi

Constructed in the mid-Qing Dynasty, Old House is in Xixinan Village of Shexian County and originally the residence of Wu Xizhi. It is the brick-and-timber construction and its first floor is lower than the second, covering an area of 342 square meters. Facing southwest, it has two sections of five-bay houses. The siheyuan, quadrangle houses, is 17.7 meters in breadth and 19.4 meters in depth. The lobby is on the ground floor of the front house, whose beam frame adopts the post and lintel construction. A crescent-shape two-step cross beam is used between the eave column and the hypostyle column, supported by T-head dougongs at both ends. The 5-purlin beam between two hypostyle columns is also supported by dougongs. On this beam there are two short columns supporting the 3-purlin beam. The downstairs central room of the last section is the living room. The gate is on the central axis. There is a pond paved with panels in the center of the yard. The front of the house is a high wall and the gate is wrapped in iron sheet under the shielding made of terrazzo bricks, which is thick and solid without any decorations. The hall upstairs is spacious and around the skywell are neat railing panels. The walls upstairs are wattled with reeds and applied with mud and lime, which are tight and fast. All the timbers in the construction are not painted in lacquer but with Chinese wood oil; so the wood grains can be seen clearly.

老屋阁位于歙县西溪南村，原为吴息之住宅，建于明代中期。为砖木结构的二层楼房，下层矮，上层高。占地面积342平方米。坐东北，朝西南，五间二进，口字形四合院，通面阔17.7米，通进深19.4米。前进楼下明间为门厅，明间梁架采用抬梁式。明间缝檐柱与金柱之间用月梁式的双步梁，两端用丁头栱承托，栱眼同住花。双步梁上用驼峰承栌斗，斗房出栱承托单步梁头。明间缝两金柱间的五架梁两端上面亦用丁头栱承托，梁上置瓜柱二，承载三架梁。瓜柱和金柱上端之间另加一单步梁。后进置瓜柱二，承载三架梁。瓜柱和金柱上端之间另加一单步梁。后进楼下明间为客厅。大门位于中轴线上，天井中央有石板砌成的水池。住宅正面为水平形高墙，大门用铁皮包镶并建有水磨砖砌成的门罩，厚实庄重，不事雕琢。楼上厅堂宽敞，沿天井四周有一圈齐整的栏板，雕有精美的飞禽走兽和花朵，还设有带扶手的"飞来椅"。楼上房壁均以芦苇编篱，表面敷泥及石灰，紧密牢固，整个建筑的木作不做大漆，只油桐油，充分显露木纹本色。

老屋阁外景 Exterior view of the Old House

中国徽派建筑　古民居

老屋阁一进天井　Skywell of the first section

老屋阁二进天井　Skywell of the second section

老屋阁三进天井　Skywell of the third section

老屋阁二进正厅 Main hall of the second section

老屋阁梁架 Beam structure

老屋阁垂莲柱头 End of drooping lotus column

巴慰祖故居 / 歙县鱼梁镇

巴慰祖故居位于歙县渔梁中街。建于清代前期。坐北朝南，建筑面积约900平方米，分前、中、后三进。前进为客厅，三楹，有天井、两庑及门厅；中、后进为住房，均为三合院，另有东、西厅。巴慰祖故居的入口与众不同，主入口是一拱门，门旁立有一块记述巴慰祖生平的石碑。柴门开在两房相邻的山墙之间，相邻的山墙间隔约有1.5米，目的是为防火。穿过门厅，为一进天井，客厅檐下和香案的上方分别悬挂着"万淑长春""敦本堂"两块匾额。梁架简朴大方，仅柱托有雕刻，三进的书房和天井之间仅有一门无墙，置画案于此，采光和景观俱佳。各院落之间均有圆门相连，别有情趣。

Former Residence of Ba Weizu

Middle Yuliang Street in Shexian County

Built in the early Qing Dynasty, Former Residence of Ba Weizu is at Middle Yuliang Street in Shexian County. Facing south, it covers an area of 900 square meters and is divided into three parts—the front section, the middle and the rear. The front section is the living-room with three principal columns, as well as a skywell, two side corridors and a hall. The middle and the rear sections are bedrooms, all being three-side compounds. There are left hall and right hall as well. The entrance of this residence is different from those of others. The main entrance is an arch with a stone stele of the biography of Ba Weizu on one side. The gate is between the gables of two contiguous houses. It is 1.5 meters wide between the two gables in case of fire. Next to the gate hall is a skywell. Under the eaves of the living-room and above the incense burner table are two inscribed boards of "Evergreen" and "Dunben Hall". The beam frame is simple and plain, and only the column bases are decorated with inscriptions. There is only a door, without wall, between the study and the skywell of the rear section. The desk is placed here, for the light and sight here are good. It is a unique taste that all the gardens are interlinked by moon gates.

巴慰祖故居正堂　The main hall

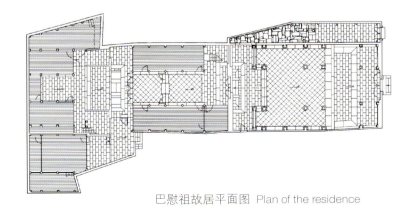

巴慰祖故居平面图　Plan of the residence

中国徽派建筑　古民居

巴慰祖故居正门 Entrance

巴慰祖故居书房　Study in the residence

巴慰祖故居二进中厅内景　Inner view of the middle hall in the second section

巴慰祖故居侧面通道(兼防火通道)　Side lane (or fire lane)

巴慰祖故居庭院　Courtyard

吴建华宅

歙县潜口村

吴建华宅位于歙县潜口村（现迁至潜口民宅博物馆），建于明代中叶早期。与司谏第毗邻，是汪善后代的住宅，近代卖给吴姓为业。据考该房初建时为3层楼房，明清间汪姓后人改为2层。该宅为砖木结构楼房，小青瓦、马头墙。面阔三间10.1米，进深10.22米，通高8米。楼层虽有改动，但楼上鹊舌式童柱、月梁、斗栱、芦苇墙、裙板及底层的梭柱、月梁、斗栱、芦苇墙、裙板及壁装修等，仍保留典型徽州明代建筑风格。正屋前面天井四周檐口装有砖制水枧槽、陶瓷落水管。由于受当时其他房屋前后左右不同条件的限制，该宅后排柱架和墙体呈斜形，不成直角；大门不在正面，而开在右廊，左廊设楼梯。

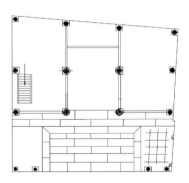
吴建华宅平面图 Plan of the residence

吴建华宅剖面图 Section view

吴建华宅窗栏 Window panel

Wu Jianhua Residence

Qiankou Village, Shexian County

Built in early period of the mid-Ming Dynasty, Wu Jianhua Residence is in Qiankou Village, Shexian County, and now it is removed to the Qiankou Museum of Folk Houses. Next to Sijiandi, it was the residence of the descendents of Wang Shan and sold to the Wus. It is said that, when it was first built, it had three stories, but the decedents of the Wangs changed it into two-storeys during the Ming and Qing Dynasties. It is the brick-and-timber construction with gray tiles and horse-head gables. The three-bay arrangement is 10.1 meters in breadth, 10.22 meters in depth and 8 meters the height. Though there are changes in stories, the magpie-tongue-like columns, the crescent beam, the dougongs, the walls and the panels upstairs, as well as the shuttle-shaped columns, the crescent beam, the dougongs, the walls and the panels downstairs still keep the features of the constructions of the Ming Dynasty in Huizhou District. Brick conduits and ceramic water pipes were installed on the cornices around the skywell in front of the main house. Restricted by other houses around at that time, the rear wall and the column frame are not in right angle but a bit tilted; the gate is not on the frontispiece but on the right corridor, and there are stairs on the left corridor.

吴建华宅正堂　The main hall

吴建华宅天井　Skywell

吴建华宅入口　Entrance

方观田宅

歙县瀹潭村

方观田宅位于歙县瀹潭村（现迁至潜口民宅博物馆），建于明中叶。三开间砖木结构楼屋，通面阔8.7米，进深7.95米，高6.65米。小砖瓦、马头墙、青砖铺地。结构紧凑、朴实，临天井有斗栱铺作托檐枋、楼板伸出额枋约一尺。尤其是屋柱与柱磉之间，加木榫，具有防震、防潮、防腐功能。隔间均采用板壁装修、芦苇墙、方格窗、合角式安榫。楼面突出檐柱额枋约1尺，承八角柱装斗栱托檐枋。大门用内外门罩保护，内门罩饰有霸王拳。为一典型的明中期普通民宅。

方观田宅外景 Exterior view

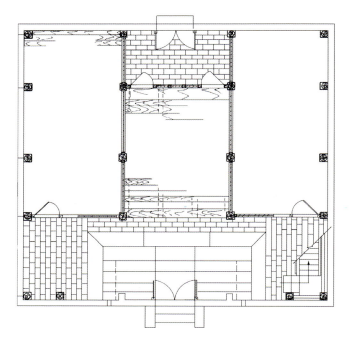

方观田宅底层平面图 Plan of the first floor

Fang Guantian Residence

Yuetan Village

Built in the mid-Ming Dynasty, Fang Guantian Residence is in Yuetan Village in Shexian County and now removed to Qiankou Museum of Folk Houses. It is the brick-and-timber three-bay construction, 8.7 meters in breadth, 7.95 meters in depth and 6.65 meters in height. It was built with small tiles, horse-head gables and gray brick-paved ground. The structure is compact and plain. Near the skywell, there are dougongs used to support architraves and the floorslabs are projecting out of the eave tiebeams. Especially the wooden tenons added between columns and column pedestals take precautions against the earthquake, humidity and corrosiveness. The rooms are partitioned with planks and have square windows. The gate is protected by the inside and outside shieldings. The inside shielding is decorated with Bawangquan. It is a typical folk residence of the mid-Ming period.

方观田宅

方观田宅正堂 The main hall

方观田宅天井 Skywell

方观田宅天井 Skywell

方文泰宅

歙县坤沙村

方文泰宅位于歙县坤沙村（现迁至潜口民宅博物馆），建于明代中叶。砖木结构，口字形四合院，二进三间楼房。通面阔9.33米，进深15.9米，高8.8米。楼下前进明间为门厅，两旁是厢房；后进明间为客厅，次间为卧室。两进之间为狭长天井，两侧为廊屋，右廊内设楼梯。楼上明间设祀祖座，楼梯口安装盖板。该房内装修非常讲究，窗外栏杆两旁望柱头雕有莲瓣，栏身上部有雕刻精美的云栱三个，下部四周嵌有雕镂精巧的镂空花板，中央用镂空方格，整个雕刻手法巧妙、技艺精湛、实乃罕见。楼面弧形栏杆是该宅最精美部分，在明间缝檐柱之间置有座板，栏杆身向外弯曲，超出檐柱外侧，形状略似椅靠背，人称飞来椅或美人靠。楼上栏杆下部裙板全部用框格式除门装饰，玲珑剔透，雕工精美。楼板之上铺楼面砖，房间四壁皮门装修。楼上房间除有望砖外，尚装有天花板。楼厅全部油漆。另外，该宅的柱础形状特别，柱础底部保持四方形，四边垂线内收，方形四角凿成下凹的弧线，上部四角斜削琢成不等边八角形，浅凹再收成圆形，图形简洁大方。为一典型的明中期富豪宅第。

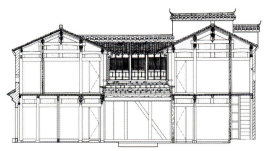

方文泰宅剖面图 Section view

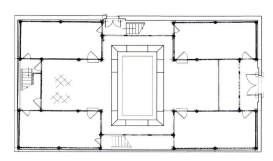

方文泰宅底层平面图 Plan of the first floor

Fang Wentai Residence

Kunsha Village

Built in the mid-Ming Dynasty, Fang Wentai Residence is in Kunsha Village, Shexian County and now removed to Qiankou Museum of Folk Houses. It is the brick-and-timber construction of siheyuan (quadrangle) and has two sections of three-bay houses, 9.33 meters in breadth, 15.9 meters in depth and 8.8 meters in height. The central room downstairs of the first section is the lobby with wing rooms on both sides; the central room of the rear section is the living room and next to it is the bedroom. There is a narrow skywell between the two sections. On both sides are the verandas and there are stairs in the right one. In the central room upstairs, there is an altar and on the stairs a cover board is installed. The fitments in this house are dainty. Lotus petals are carved on the capital of the balusters beside the railings outside the windows. There are also carvings on the upper part of the railings. On the lower part are embedded panels of exquisite ornamental engravings. In the center are the through-carved panels. All the sculptures are superb in technique and exquisite in craftsmanship, which is unique. The arc railings on the floor are the most exquisite of the whole house. The railings bend outside beyond the outer part of the peripheral columns. Its shape is like the back of a chair and so it is also called beauty chairs or flying chairs. The panels on the lower part of the railings upstairs are delicately decorated. The floor is paved with bricks. The rooms upstairs have ceilings in addition to the decoration of watching bricks. The balcony is painted in oil. Besides, the column bases are peculiar in square shape but cut into round or arc corners. This is a typical residence of the rich in the mid-Ming Dynasty.

方文泰宅天井 Skywell

中国徽派建筑　古民居

方文泰宅二层　Second floor

方文泰宅正堂　The main hall

方文泰宅正门入口 The entryway

方文泰宅木雕 Wood carvings

德庆堂

歙县琶塘村

德庆堂位于歙县西溪南琶塘村（现迁至潜口民宅博物馆），建于明代中叶。此宅的最大的特点是结构与众不同，不仅上、下两层的柱网不对齐，而且以屋脊为界前后开间亦不同。脊前、脊后分成两进，脊前五开间，其中中间为明堂，堂上高悬"德庆堂"的匾额，两侧的梢间为卧室，脊后分成四间，两间相套，形成两个套间。二层的客厅宽敞明亮，两侧有厢房，楼上和楼下均为方砖铺地，得庆堂的雕刻也非常精美，围绕天井三面的楼行上，雕满了花、鸟、鱼、虫和动物图案，雕刻手法有平雕、透雕、镂空雕和高浮雕，充分展示徽州工匠的高超技艺。

Deqing Hall

Patang Village, Shexian County

Built in the mid-Ming Dynasty, Deqing Hall is in Patang Village, Shexian County, Shexian County and now removed to Qiankou Museum of Folk Houses. The most unique characteristic of this residence is its peculiarity in structure. Columns of the two storeys are not paralleled; moreover, the bays of the two parts divided by the ridge are not the same. The front part has five bays, with the central room as the lobby above which there is an inscribed board of "Deqing Hall", and the side rooms are bedrooms. The rear part is divided into four bays forming two flats. The living room upstairs is bright and spacious with two wing rooms. All the floors are paved with bricks. And the sculptures of this hall are very exquisite—at the corridors upstairs around the skywell are full of inscriptions of flowers, birds, fishes, insects and animals; the techniques include flat carving, openwork carving, through-carved work and alto-relievo. All these reflect the superb workmanship of the artisan in Huizhou District.

德庆堂天井 Skywell

德庆堂二层裙板　Panels on the second floor

德庆堂全貌　A full view of Deqing Hall

德庆堂正堂　The main hall

德庆堂二层正厅 The main hall on the second floor

遵训堂

歙县棠樾村

遵训堂位于歙县棠樾村，建于清嘉庆年间，为鲍志道弟鲍启运之私宅，与保艾堂门相向。正房毁于咸丰太平天国之役，现仅存东侧隔一弄的"存养山房"和后进的"欣所遇斋"。两处均为厅室，用一面极大的花墙相隔，昔为会客、宴庆、佣工事务操持所。"存养山房"匾额为清王文治所书(匾存村中)，"欣所遇斋"四字篆书为胡长庚书。

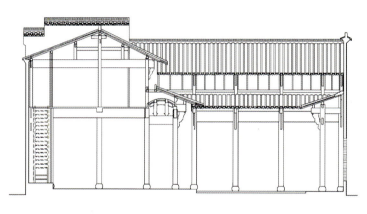

遵训堂剖面图 Section view

遵训堂屋内 Inside the Zunxun Hall

Zunxun Hall
—The Hall of Following Teachings
Tangyue Village of Shexian County

Constructed in the Jiaqing reign of the Qing Dynasty, Zunxun Hall is in Tangyue Village of Shexian County. It was the private house of Bao Qiyun, Bao Zhidao's brother, opposite to Bao'ai Hall. The main house was destroyed in the war against the Kingdom of Heavenly Peace in the reign of Xianfeng, and only Cunyang Shanfang (a villa) across a lane on the east and Xinsuoyu Zhai (Hall of Happy Gathering) in the rear section are left. These two halls, both separated by a huge lattice window, were used to meet guests, hold banquets and also for servants' use. The inscribed board of "Cunyang Shanfang" was written by Wang Wenzhi in the Qing period and now is kept in the village. The board of "Xinsuoyu Zhai" was written by Hu Changgen in seal characters.

遵训堂

存养山房入口 The entryway of the main hall

存养山房天井大厅 The main hall and the yard

欣所遇斋正堂 The main hall of Xinsuoyu Zhai

存养山房花墙　The fretted wall

存养山房正厅　The main hall of Cunyang Shanfang

保艾堂

歙县棠樾村

保艾堂位于歙县棠樾村，由当时的巨贾两淮盐务总商鲍志道父子建于清嘉庆初年。《诗经·小雅·南有台》篇"乐只君子，保艾尔后"，《礼记》"五十曰艾"，盖堂主人之前辈，多有夭寿者而名以祈保焉。该宅是古徽州最大的一栋宅第，因信奉道教，故朝北向。建成时有一百零八间房间，三十六个天井。按八卦方位避凶就吉，组合而成。入院大门，内有门厅及廊房一列；进而为正式大门，建有精致的砖雕门罩；再入为保艾堂主厅外之廊屋，一条串通大屋前后（南北方向）的避弄巷，建筑分为东、西两部分，最后是花园。东部包括三个装饰华丽的厅，即白果厅（悬挂保艾堂匾之正厅，现存），楠木厅（正厅右之花厅，有"安且吉兮"匾额，已毁），红木厅（花厅后之倒厅，早年毁于火）。另有"安素轩"书斋，因志道父子集刻"安素轩法帖"，闻名于世。后进作大灶房，再后作十间仓屋，墙内上下左右均置有防鼠铁丝网。保艾堂大屋设计建造精严，一般上廊有二层楼房，下廊则是平屋。每进房屋均有高大封火墙。地面铺设也非常讲究：一层石灰、一层细砂、一层酒缸(口朝下覆盖排列在底层)，上面再加一层砂、最后才铺上地墁砖。因此，即使倒水在地上，也很快会被吸干，遇到梅雨季节从来不返潮，干燥爽朗。全屋安装的檐沟、落水管，均选用优质铜锡铸成，门栓、环悉用铜。据说大屋全部木材均采自四川，工匠则请自扬州。

Bao'ai Hall

Tangyue Village, Shexian County

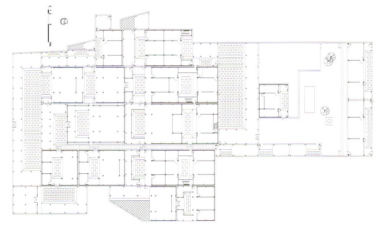

保艾堂平面图 Plan of Bao'ai Hall

Bao'ai Hall was built by Bao Zhidao and his son, rich merchants and chief businessmen of salt affairs in the area of the Huaihe River at the time, in the early Jiaqing reign in the Qing period. The owner of the house named it "Bao'ai" (quoted from an ancient poem in the Book of Songs) for praying longevity, because many of his ancestors had died young. This hall is the largest of the ancient houses in Huizhou District. It faces north in belief of Taoism. When it was first built, there were 108 rooms and 36 skywells, an arrangement to avoid impending troubles and seek good luck according to Bagua (the Eight Trigrams). Inside the gate, there are a lobby and a row of veranda; then there is the formal gate, above which is the refined gate-shielding of brick carving; and then there are the veranda outside the main hall and a north-southward alley through the house. The house is divided into two parts—east section and west section; the last part is the garden. There are three gorgeous halls in the east part—Baiguo Hall (Hall of Ginkgo, the main hall where the board of Bao'ai Hall is hung and still exists), Nanmu Hall (the parlor to the right of the main hall, with the board of "Peace and Happiness" which was destroyed), and Padauk Hall (the reversible hall behind the parlor, which was damaged by fire in the early years). Besides, there is the study of "Ansuxian", which is famous for its model calligraphy collected and engraved by Bao Zhidao and his son. The main hall is dainty in the design and construction. There are high gables in every section and the ground is tastefully paved. All the eave gutters and pipes are made of copper and tin of high grade, and the door bolts and knockers are made of copper. It is said that all the timbers were from Sichuan and all the craftsmen were from Yangzhou.

保艾堂天井 Skywell

保艾堂入口 Entrance of Bao'ai Hall

保艾堂白果厅入口 Entryway of Baiguo Hall

保艾堂避弄巷 Alley of Bao'ai Hall

The Stage of Wang Clan

—

County Town of Xiuning

汪氏家戏台 ／ 休宁县县城

The Great Stage of Wang Clan is in the county of Xiuning. Facing south, it was built in the late Ming Dynasty. The plan arrangement is not symmetrical. The gate is five meters behind the gables on both sides to make a grain-drying area. The gate is an arch with a blackstone stele on which are the two Chinese characters of "Lu Yuan". Behind the gate is a skywell on the right and there is a square pond in the skywell, with railing panels around. Next to the pond is the family stage of three run-through bays covering an area of 100 square meters. The right side of the stage is connected with the main hall, on the second floor of which are the living room and wing rooms. There are down-to-floor partition doors upstairs along the skywell with railings outside, where people could watch plays. On the left side of the skywell are the two-story building of the wing rooms, the study and the kitchen. The outermost room has a door open to the outside.

汪氏家戏台位于休宁县县城。建于明代后期，坐北朝南。平面布局不对称，大门内退两边山墙5米，留出一个晒场，门为拱门，上有一块青石匾额，刻"卤园"二字，进门后天井在左侧，天井中有一方形水池，四周有栏板，水池后是家庭戏台，戏台三开间贯通，有100平方米，与戏台右侧相连的是正堂，二楼有客厅和厢房，二楼对天井一侧均为落地格扇门，外有栏杆，亦可在二楼观戏。天井的左侧是厢房、书房、厨房等，为二层楼。底层最外一间有一门对外。

汪氏家戏台入口 The entryway of the stage

汪氏家戏台全貌 A full view of the stage

汪氏家戏台天井及戏台 The skywell and the stage

汪氏家戏台

程大位故居 / 休宁县屯溪镇

程大位故居位于屯溪区上新率口渠东。两进三间砖木结构二层楼房。建于明正德年间。占地500平方米，门楼里外挑檐、曲梁斗栱，马头山墙，围墙的漏窗亦做成算盘状，别有一番情趣。室内气氛恬静，陈设简朴。现辟为程大位纪念馆，第一进西房为程大位原居室。

程大位故居院落门口　The entryway of the courtyard

程大位故居围墙漏花窗　The fretted window on the wall

Former Residence of Cheng Dawei

Tunxi Town, Xiuning County

Former Residence of Cheng Dawei is to the east of Shuaikou Canal in Shangxin, Tunxi District. It is the brick-and-timber construction with two sections of two-story three-bay houses. Built in the Zhengde reign of the Ming Dynasty, it covers an area of 500 square meters. The gate tower has the projecting eaves inside and outside. It adopted curved beam, dougong and corbie gables. It is interesting that the fretted windows on the bounding walls were made in the shape of abacus. The hall has a tranquil atmosphere and is simply furnished. Now it is Museum of Cheng Dawei, whose former bedroom was the west room of its first section.

程大位故居

程大位故居入口 The entrance

程大位故居正堂 The main hall

程氏三宅 / 休宁县屯溪镇

程氏三宅分别坐落于屯溪柏树街东里巷6、7、28号。明成化年间礼部右侍郎程敏政所建。二宅均为封闭式砖木结构两层楼房，分前后两进，俗称"一脊两堂"，屋面盖蝴蝶瓦，四周墙体封护，东、西马头山墙，前后檐墙砌凹形。6号宅为五开间，占地面积477平方米，7号、28号宅为三开间，占地面积分别为154平方米和187平方米。为增大堂屋开间，三宅底层均用抬梁式构架，浑厚月梁穿入金柱，梁柱间以丁头栱承托，栱眼雕花，柱下置八角形石柱础，础面垫一块5厘米厚木质，木质唇口与柱础石面相吻合。

底层次间、梢间的棂窗障目板素装，槛窗棂制成斜方格，左右配以雕刻图案。活动窗扇边呈拱形，格心制方格眼，既遮挡视线又拆装方便，具有疏朗空透之效果。

楼层井檐上装活动排窗，外挑的垂莲柱四方抹角，内侧装置飞来椅，椅脚为豹爪形，外侧上端插拱出两跳，托住檐檩。支撑外挑垂莲柱的斜撑为鹅颈状，下端插入金柱，形成适度的传力结构。

三宅内灵芝如意状井檐斜撑，用透、剔、掏、挖等手法雕六朵层次迭落的灵芝卷瓣，使整个装饰物形成六面观看的花罩，同时又将卷瓣的起突面分层次雕刻，使卷瓣脉络清晰，完全立体化。这种雕饰，既继承宋"营造法式"的技巧，又糅合明代透雕风格，是徽州宅第木雕装饰的典型代表。

7号宅第内的井檐裙板，自下而上五道花板横向组装，同时又以直装灵芝卷草及荷叶纹板相隔，使整个花板裙面变化突出，生动活泼。

Three Houses of the Cheng Clan

Tunxi Town, Xiuning County

Three Houses of the Cheng Clan are located respectively at No. 6, No. 7, and No. 28 of Dongli Lane, Baishu Street in Tunxi. They were constructed by Cheng Mingzheng, vice minister of Ministry of Rites, in the Chenghua reign of the Ming Dynasty. All of the three houses are close-type brick-and-timber two-story buildings, divided into two sections ("two halls of one ridge" in local name). The roof is covered with Chinese convex and concave tiles and the walls around are also covered and protected. On the east and the west are corbie gables and the front and rear eave walls are of the concave shape. The house at No. 6 is of five-bay style and the other two at No. 7 and No. 28 are of three-bay style. In order to enlarge the room of the hall, the first floors of the three houses are of the post and lintel construction. The simple and vigorous crescent beam crosses into the principal column. The beams are supported by T-shape carved Dougongs on the columns. Under the columns are the octagonal stone bases with underlays of five-centimeter thick planks which have the grains similar to the veins of the bases.

There are blind eye-blocks on the window lattices of the subcentral and end bays on the first floor. And the sill window lattices are made into oblique panes with engravings on both sides. The edges of the movable sashes are arch-shape and in

程氏三宅天井 Skywell

五道花板的底层雕瑞兽，其形态栩栩如生。同时瑞兽两侧各雕四乳钉旋转水浪花牙，海水成条浪状向中间翻涌，组成海水江牙图案。

第二道花雕番莲卷草，纹样以半坡状线条与切圆组成二方连续的装饰带，并在切圆中间雕宝相花，旋花花瓣向四周伸延翻卷。同时又在波线上加子叶，构成花叶繁盛茂密、枝莲层层叠压、卷花穿枝过梗、旋花弯转缠绕、疏密有致的缠枝莲。

第三道花板雕国色天香牡丹和长寿雷菊，花卉或盛开怒放或含苞待放。花丛中的喜鹊、翁鸟栖枝梗，动作有回首转视、仰颈天空、闪翅鸣啼、挠尾欲飞等多种形态。匠人以剔地起突的技巧，突出禽鸟、花卉和叶面，并在它的背面留数株小圆柱支撑花鸟，视之有镂空之感。

第四道花板雕灵芝如意图案，灵芝剔地，周留粗犷线条，使画面形成凹、平、凸的式样，图案脉理清晰，线条通蒙大方。

顶端雕荷花状斗栱，荷叶面向外翻卷，呈多层次重叠，舒展得宜，从而产生叶面浮起的真实感。

7号、28号宅第木构架不加髹漆，木纹显露，古朴大方。6号宅第底层部分木柱，门用麻织物粘贴底纹，然后加油灰刷色。明间仰顶绘旋子画，上、下层大梁通绘包袱画和角叶彩画，填彩以冷色为主。

28号宅第门罩采用麻石料凿制，仿木结构四柱三楼式，其正楼和次楼均以斗栱相托。石制上额枋雕飞凤戏牡丹，下额枋雕双狮抢球，平板枋雕串莲花瓣。6号宅第里门楼砖雕，额枋上雕满地翻叶大花。

6号宅第存有明天启元年（1621年）买卖房契一份，楼层太师壁门留有徽派明代木刻填彩年画一张，人物形象清晰可辨。

the centers are checkered patterns to shelter from sight or to remove easily; thus it has a bright and clear effect.

A row of movable windows are installed along the eaves upstairs around the skywell. The projecting drooping lotus columns have round-off corners. There are flying chairs which have leopard-claw-like feet. The upper outside of the plugged bracket arms extend for two tiao to hold eaves and purlins. The diagonal braces holding the projecting drooping lotus columns are gooseneck-like and their ends penetrate into the principal columns. That makes a proper transmission frame.

In the three houses, the diagonal braces under the eaves around the skywell are inscribed into six crockers of glossy ganoderma petals by means of openwork carving, cut-out carving, undercutting and excavating. It makes the ornament look like a flower-cover from six perspectives and the raised petals are engraved by hierarchical structure to make the venation distinct and the petals stereoscopic. This kind of carving is typical in the decoration of wood carvings in the folk houses of Huizhou, combining the skills in Building Formulas in the Song Dynasty and the fretwork of the Ming Dynasty.

程氏三宅28号入口 Entrance of No. 28

程氏三宅6、7号前院 Front courtyard of No. 6 and No. 7

程氏三宅"三保险"锁 Door-locks

程氏三宅6号门楼内侧石雕 Stone carving above the back of the gate

程氏三宅二层裙板木雕 Carved panels on the second floor

程氏三宅6号二楼 The second floor of No. 6

中国徽派建筑　古民居

程氏三宅6号正堂　The main hall of No. 6

程氏三宅7号正堂　The main hall of No. 7

程氏三宅

程氏三宅天井 Skywell

金馀庆堂 \ 婺源县延村

Jin Yuqing Hall

Yancun Village, Wuyuan County

    Built in the thirteen year of the Kangxi reign of the Qing period, it is located in Yancun Village in the middle part of Wuyuan County. Its gate was built with terrazzo bricks and the shielding has petals and angles and upturned eaves. The structure of the gate appears like the shape of Chinese character of "商"(business or businessman), and it has the meaning that it was built with the money earned from business. The gate is on the left corner of the house and the gatehouse used to be the place of parking the sedan chair of the guest. Fine engravings are furnished on the girder and tie beams, queti, window lattices, partition boards and the lintel of the main hall. The panelings of the left and right wing rooms are inscribed with designs of "Happiness, high position and long life", "The Kylin Bestowing a Boy", "Three Goats", and "Pines and Cranes". On the front beam is the design of bats and on the left and right side beams are the designs of ocean waves; all these combined have the meaning of "Happiness as immense as the East Sea". Below the beam of the main hall is the sculpture of "Two Phoenixes Facing the Sun". And on the side cross tie beams are the characters of " Shou Bi Nan Shan" (meaning "Long life as Mountain Tai") and "Tian Guan Ci Fu" (meaning "Blessings from the Heaven"). These inscriptions, which have profound implications and meaningful messages, not only embody the superb skill of the craftsmen in Huizhou, but also reflect the people's sweet dreams of happy life. On the stone base at the rear skywell, at the inlet of the sewer, is the engraving of a propitious Kylin. Three of its feet are stepping on three copper coins and another coin under the fourth foot is on the black stone panel beside the floor drain. This is the only design the author has ever seen in ancient Huizhou architecture.

金馀庆堂全景　Overall view of Jin Yuqin Hall

金馀庆堂位于婺源县中部延村，建于清康熙十三年。石库门坊、水磨清砖门面，门罩翘角飞檐，结构似一个"商"字，意为主人当年经商致富而建。金馀庆堂的院门开在房屋的左角，进门有一门屋，古时用于停放来宾的轿子。正堂梁枋、雀替、窗棂、隔扇、门楣等都有精心雕刻，左、右厢门芯板雕有"福禄寿喜"、"麒麟送子"、"三羊开泰"、"松鹤延年"等图案，前梁上刻有蝙蝠图案，与左右两侧梁上的海浪图案合为"福如东海"。正堂梁下雕有"双凤朝阳"。两侧横枋上则刻有"寿比南山"、"天官赐福"。这一幅幅含意隽永、寓意深刻的精美木雕，不仅充分体现了徽州工匠高超的雕刻技艺，亦反映了人们对生活的美好憧憬。值得一提的是，在后院天井的护石上，在下水口处雕有象征着吉祥的"麒麟"图案，麒麟三只脚分别踏着三枚铜钱，另一只脚下的铜钱则刻在地漏旁的青石板上，笔者在古徽州仅见此一处。

中国徽派建筑　古民居

金馀庆堂大门　The entrance

金馀庆堂后天井　The back skywell

金馀庆堂正堂　The main hall

金馀庆堂

金馀庆堂绦环板木雕 Carvings on the panels

梁下"双凤朝阳" Wood carvings under the beam

聪听堂

婺源县延村

聪听堂位于婺源县中部延村，建于清康熙年间。两进三开间，入口有前院，前院右边是院门，左边是柴门。正中是"聪听堂"的大门，水磨清砖门面，门罩翘角飞檐，门头上的砖雕讲究。入门正堂，典型的徽州民居布局，梁枋和两厢的雕刻精美，特别是两厢隔扇门的芯板刻着唐代大诗人白居易"琵琶行"中"浔阳江头夜送客，枫叶荻花秋瑟瑟"、"千呼万唤始出来，犹抱琵琶半遮面"四句诗的写意图案。右侧的阁板上雕刻的八个"花瓶"，更是惟妙惟肖，令人称绝。

聪听堂天井 Skywell

Congting Hall--Hall of Sharp Hearing

Yancun Village, Wuyuan County

Built in the Kangxi reign in the Qing Dynasty, it is in Yancun Village in the middle part of Wuyuan County. There are two sections of three-bay houses and next to the gate is the front yard. On the right is the gate leading to the yard and on the left is the side door. In the middle is the main entrance with the surface covered by terrazzo bricks. The gate shielding has the petals and angles and upturned eaves. The brick carvings on the door head are dainty and pretty. Behind the gate is the main hall. It has a typical arrangement of folk houses in Huizhou. The engravings are exquisite on the girders and tie beams, and especially in the wing rooms, on the panels of the partition doors are the designs of artistic conception of the well-known verses by Bai Juyi, a famous poet of Tang Dynasty. The eight vases inscribed on the right partition boards are admirably vivid and lifelike.

聪听堂

聪听堂入口　The entryway

中国徽派建筑　古民居

聪听堂正堂　The main hall

聪听堂

千呼万唤始出来，犹抱琵琶半遮面 Wood carving of the image of a Tang poem

浔阳江头夜送客，枫叶荻花秋瑟瑟 Wood carving of the image of a Tang poem

聪听堂匾额 The inscribed board

聪听堂隔扇门木雕(局部大样) Details of the wood carvings on the partition doors

聪听堂隔扇门木雕(局部大样) Details of the wood carvings on the partition doors

云溪别墅

婺源县理坑村

Villa of Yunxi

Likeng, Tuochuan
of Wuyuan County

　　The Villa of Yunxi is in Likeng, Tuochuan of Wuyuan County. It was built in the Daoguang reign of the Qing period by Yu Qiguan (an intimate imperial aide and adviser). The ancient villa was named after Yu Qiguan, whose literary name was Yunxi. On the architrave of the gate are the characters of "Yun Xi Bie Shu". There are drooping lotus pedestals on two sides. The gate is a wood multi-eave tower supported by a pair of qiangjiao extended with eight sets of dougongs. On the architrave inside the gate tower is a rare design of "Fairy Cranes". The cranes have different shapes and spirits and are vividly portrayed. There are two rows of beauty chairs on both sides inside the gate. Behind the gate is a large yard. The main hall is to the right of the gate tower, which is of one section, five-bays with two wing rooms beside. Under the front eaves are round ridge roofs. On the whole architrave is a sculpture of "Two Phoenixes Facing the Sun". Opposite to the gate tower is the study of the owner, which is small but quiet.

云溪别墅入口倒座 The reverse side of the entrance

云溪别墅大门 The entrance

"云溪别墅"位于婺源县沱川理坑，建于清道光年间，系道光年间光禄大夫余启官所建。余启官，号云溪，古别墅名为"云溪别墅"。大门的额枋上书有"云溪别墅"四个大字，两侧垂莲柱。大门内为木结构重檐门楼，八组斗栱挑出两对戗角，支撑着整个门楼。门楼内的额枋上雕有一组难得一见的"仙鹤图"，图中的仙鹤个个神形兼备，有呼之欲出之感。门内两侧各置有一排"美人靠"，供人歇息。门后是一个大的院落，正厅在门楼右边，一进五开间，左右有厢房，前檐下有卷棚。整个额枋雕刻着一幅"双凤朝阳"。门楼的对面是主人的书房，书房小巧清静

云溪别墅边门入口　The entryway of the side door

云溪别墅"九世同堂"木雕 Wood carving of "Nine-Generations Living Together"

云溪别墅"双凤戏牡丹"木雕 Wood carving of "Two phoenixes and peonies"

云溪别墅门上木雕"仙鹤图" Wood carving of "Fairy Cranes" above the gate

俞氏客馆

婺源县思溪村

俞氏客馆位于婺源县思口镇思溪村。建于清乾隆年间，二进三开间，入口前院不足二米，特别是在十扇隔扇门的绦环板上阳刻有九十六个不同字体的"寿"字，其中四块刻有十二个"寿"字，六块刻有八个"寿"字，另外还有四个"寿"字，一个在中间门的横板上，有两个在内厢房的窗扇上，最后一个"寿"字不知在何处。以上共同组成"百寿图"。

俞氏客馆"百寿图"木雕 Wood carving of "One Hundred Characters of Longevity"

Guest House of the Yu Clan

Sixi Village, Wuyuan County

Built in the Qianlong reign of the Qing Dynasty, it is in Sixi Village, Sikou of Wuyuan County. It has two sections of three-bay houses. The front yard at the entrance is less than two meters in depth. On the panels of the ten partition doors are relieves of ninety-six Chinese character "Longevity" in different styles of calligraphy, among which four doors have 12 characters for each and six have 8 characters each. About the other four characters, one of them can be found on the cross panel of the middle door and two on the window sashes in a wing room, but the last one is hidden somewhere. And all this makes a design of "One Hundred of Longevity".

俞氏客馆

俞氏客馆入口 The entrance

俞氏客馆隔扇门 Partition doors

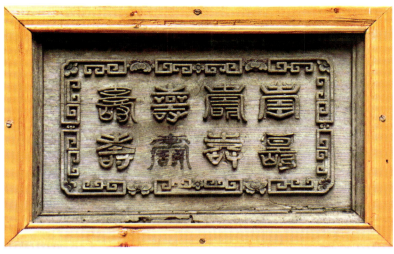

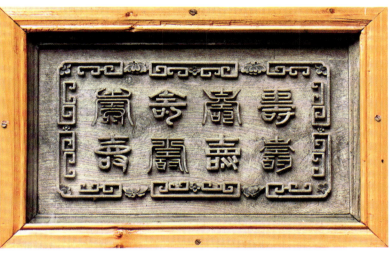

俞氏客馆"百寿图"绦环板 Panel carvings of "One Hundred Characters of Longevity"

胡寿基宅

绩溪县上庄村

胡寿基宅坐落在绩溪县上庄镇上庄村适之路25号。坐北朝南，砖木结构，硬山屋顶，抬梁穿斗并用式梁架，水磨方砖地坪。面阔五间，进深七间，占地面积234平方米。该宅分前后两堂，中设天井。大门前有一庭院，大门及前檐墙窗户均用水磨青砖砌筑并雕饰细腻图案。大门内设木质屏风墙壁。前堂及两厢撑栱圆雕人物，后堂撑栱圆雕雄狮。二楼天井四周有短窗装修。其窗户、隔扇、雀、枋盖板均有繁复的雕刻画面。尤其是隔扇的装饰题材颇具文化品位，中绦环板高浮雕镌刻着人物故事，裙板则雕刻文字诠释人物故事，令人一目了然。该宅保存完好，虽为民国初年建筑，但确有较高的艺术价值。

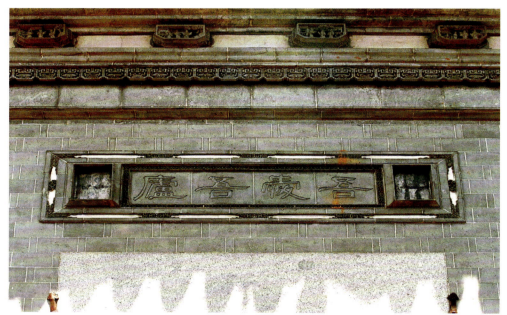

胡寿基宅"吾爱吾庐"门匾额 The inscribed board of "I Love My House" above the gate

Hu Shouji Residence

Shangzhuang Village, Shangzhuang Town of Jixi County

Hu Shouji Residence is located at No. 25, Shizhi Road in Shangzhuang Village, Shangzhuang Town of Jixi County. Facing south, the house is the brick-and-timber construction with flush gable roof. It adopts the combination of post and lintel construction and through-jointed frame. The ground is paved with terrazzo bricks, measuring five-bay in breadth and seven-bay in depth and covering an area of 234 square meters. The house is divided into two parts with a skywell in the middle. There is a yard in front of the gate. The gate and the windows of the front wall are laid with terrazzo bricks and decorated with delicate inscriptions. Inside the gate is a wooden screen wall. The arch struts in the front hall and in the wing rooms are furnished with medallions of characters and those in the rear hall with medallions of lions. Small windows were installed around the skywell on the second floor. Heavy and complicated designs were inscribed on the windows, partition boards, queti, and tie-beam cover-boards. The theme of the decorations on the partition boards is especially of high taste. Stories of characters were carved in high relief on the middle taohuan boards and on the panels are carved the explanation of the stories to make them clear. This house is in good preservation. Though built in the early years of the Republic of China, it still has high cultural value.

胡寿基宅入口 The entryway

胡寿基宅正堂 The main hall

胡寿基宅二层　The second floor

胡寿基宅二层窗扇　Windows on the second floor

胡寿基宅鲤鱼跳龙门灯钩　The lantern hook

胡寿基宅厢房窗户 Windows of the wing room

胡寿基宅隔扇门头 Head of the partition doors

胡寿基宅隔扇门 Partition doors

胡适故居

绩溪县上庄村

胡适故居位于绩溪县上庄村。建于1877年,为胡适的父亲胡铁花所造。故居围有墙院,楼在院中,坐北朝南,两进一楼通转式结构。门首嵌着四块古雅精致的砖雕,上有飞檐翘角的门罩。前进是正厅,东西两间卧室各连厢房。卧室窗门栏板和两厢房12扇落地槛门上的木雕为墨模雕刻能手胡国宾所雕。西边卧室是胡适与江冬秀结婚的洞房,连着西厢的是胡适幼年读书处。大厅前和正厅上各悬着当代著名书法家沙孟海题写的"胡适故居"直、横两块金字牌匾。正厅当中挂有胡适画像。故居的摆设如旧,陈列有胡适结婚时所睡的"月宫床"、胡铁花当年佩用的七星宝剑、胡适家书真迹复制品等。

Former Residence of Dr. Hu Shi

Shangzhung Village, Jixi County

The Former Residence of Dr. Hu Shi is located in Shangzhung Village, Jixi County. It was built by Hu Tiehua, Hu Shi's father in 1877. Facing south, the house is in a courtyard surrounded by the enclosing walls and it is of two-section structure with the first floor passing through. On the door head are embedded four classic elegant brick carvings and over it is a shielding with upturned eaves. The main hall is in the front section and the left and right bedrooms are connected with wing rooms. The wood carvings on the frieze panels of the windows in the bedroom and those on the 12 partition doors in the wing rooms were inscribed by Hu Guobin, the master-hand of ink-sculpture at the time. The west bedroom was the bridal chamber for Hu Shi and Jiang Dongxiu. The next room connecting the west wing room was the study of Hu Shi when he was young. A horizontal gilded plaque and a vertical one are hung in front of the lobby and in the main hall respectively, written by Sha Menghai, the contemporary famous calligrapher. A portrait of Hu Shi is hung in the middle of the main hall. The furniture is placed as before, including "the Bed of the Moon Palace" used by Hu Shi when he married, the seven-star sword worn by Hu Tiehua and the copies of Hu Shi's family letters.

胡适故居

胡适故居入口　The entrance

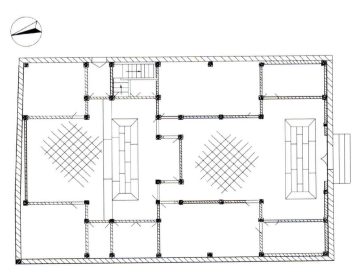

胡适故居底层平面图　Plan of the first floor

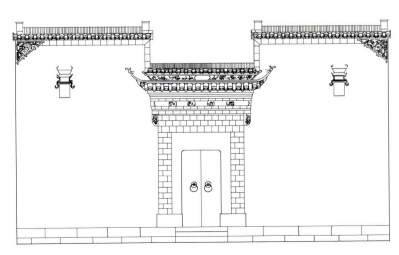

胡适故居正立面图　Facade of the residence

中国徽派建筑　古民居

胡适故居正堂　The main hall

胡适故居

胡开文纪念馆

绩溪县上庄村

胡开文纪念馆坐落在绩溪县上庄镇上庄村东南首。胡开文，名天注，字柱臣，是清代四大制墨名家之一，纪念馆坐西朝东，砖木结构，硬山屋顶，山墙砌筑封火墙，抬梁、穿斗并用式梁架，面阔三间，进深六间，占地面积126平方米。撑栱、雀替、窗户、隔扇均饰雕刻。两幢建筑均未维修过，呈清代末年民居建筑原有风貌。

胡开文纪念馆入口 The entrance

Memorial Museum of Hu Kaiwen

Shangzhuang Village, Jixi County

The Memorial Museum of Hu Kaiwen is on the southeast of Shangzhuang Village in Shangzhuang Town, Jixi County. Hu Kaiwen, with an assumed name of Tianzhu and styling himself as Zhucheng, was one of the four famous experts in ink-making in the Qing period. Facing east, the museum is the brick-and-timber construction with yingshan—flush gable roof. The walls were built in fire-sealing gables. The post and lintel construction is adopted as well as the through-jointed frame. The museum measures three-bay in breadth and six-bay in depth and covers an area of 126 square meters. There are carvings on the arch struts, queti, windows and partition boards. Both of the two houses have never been repaired and still remain the style and features of the folk houses at the end of the Qing Dynasty.

胡开文纪念馆

胡开文故居正堂 The main hall

胡开文纪念馆后坐 The reverse side

石磐安宅

绩溪县石家村

石磐安宅坐落在绩溪县上庄镇石家村35号。清代晚期建筑。坐南朝北，硬山屋顶，砌封火山墙，抬梁、穿斗并用式梁架，方砖地坪。南北两堂相对，中设天井，如意花卉撑栱承挑出檐，檐部四向有矮窗装置，二楼通转。面阔3间10.3米，进深8间12.3米，占地面积126.29平方米。大门开在北檐墙中部，门框上方是五飞砖门罩，砖雕完好精致。大门内设木板影壁。其后的石质天井，石框内有深28厘米的水池。内装饰构件有：狮形、人物撑栱各4件，两厢檐部隔扇8件，花心是纹嵌人物故事，裙板为素面；窗门4件，花心亦为纹嵌博古，4件遮羞板饰深浮雕山水人物图，技艺精湛，具有较高的文物价值。未经修缮，保存完好。

Shi Pan'an Residence

Shijia Village, Shangzhuang Town of Jixi County

石磐安宅倒爬狮斜撑 A lion-shaped bracket

Built in the late Qing Dynasty, Shi Pan'an Residence is located at No. 35 in Shijia Village, Shangzhuang Town of Jixi County. Facing north, it adopted the post and lintel construction as well as the through-jointed frame with flush gable roof. The walls were built in fire-sealing gables and the ground was paved with quadrels. The north hall is opposite to the south hall with a skywell in the middle. The ruyi and flower shape brackets extend to support the eaves, around which the dwarf windows were installed. The second floor can be passed through. It is three-bay of 10.3 meters in breadth and eight-bay of 12.3 meters in depth, covering an area of 126.29 square meters. The main gate is in the middle of the north eave wall. Above the doorframe is a brick shielding and the refined brick carvings are in tact. In the gate is a wood screen wall. Behind the wall is a stone skywell. In the stone case is a pond about 28 centimeters deep. The interior decorative elements of the house include four lion-like arch struts, four man-like arch struts, eight partition boards of the wing room eaves. The huaxin of the boards is the design of characters and the panels are plain. There are also four windows, the huaxin of which are embedded with curios, and four coverings inscribed with deep relieves of people and mountains and waters. The skills were consummate and all of these have high historic value. They have never been repaired but are still in good condition.

石磬安宅香案 The incense burner table

石磬安宅二层 The second floor

石磐安宅正堂 The main hall

跑马楼

黟县西递村

跑马楼位于黟县西递村村口，又名"凌云阁"，建于清乾隆丁未年（1787年），系明经胡氏二十四世主、江南六大首富之一、正三品、通议大夫胡贯三为了迎接三朝元老宰相亲家曹振镛而建。跑马楼为二层，由一个跑马廊和亭子组成，底层外围封闭，但开有数十个漏花窗，亭为歇山顶，粉墙黛瓦，檐角悬挂风铃，跑马廊宽3米，长约28米，两侧有护栏，院内还种有数十棵桃树。登上跑马楼二层，漫步在眺廊上，凭栏远望，青山绿水，尽收眼底，阵阵清香的山风吹来，令人心旷神怡。

跑马楼内院 The inner courtyard

跑马楼外景 The exterior view

Paoma Tower–Horse Racing Tower

Xidi Village in Yixian County

Horse Racing Tower, also called "Cloud Reaching Pavilion", is at the entrance of Xidi Village in Yixian County. To welcome Cao Zhenyong, a prime minister of the court who had served under three emperors in a row and also his relative by marriage, the tower was built in the Qianlong reign of the Qing Dynasty (1787) by Hu Guansan, the descendent of the 24th generation of the Hu Clan, one of the six richest in southern area of the Yangtze River and the third ranking official of the court. The Horse Racing Tower has two stories and consists of the Horse Racing Porch and a pavilion. The first floor is enclosed outside but there are tens of ornamental lattice windows. The pavilion was built with the xieshan roof, white-washed walls and gray tiles. Under the eaves are hung wind bells. The Horse Racing Porch is 3 meters wide and 28 meters long. On both sides are the railings. In the yard, there are tens of peach trees. On the porch upstairs, you can have a panoramic view of the mountains and waters and feel carefree and joyous in the wind blown from the mountains.

跑马楼

跑马楼二层眺廊 The viewing gallery on the second floor

西递村入口—跑马楼全景 Entryway of Xidi Village—Overall view of the Horse-racing Tower

跑马楼

瑞玉庭

黟县西递村

瑞玉庭位于黟县西递村，建于清咸丰三年，即公元1853年，房子的原来主人是徽商的佼佼者。庭院里的门罩上左右两边各有一块元宝式的砖雕刻有"富、贵"二字，其意就是希望自己招财进宝，大富大贵；在院子的右墙上还刻有"履道含和"四个大字，其意就是说，在人生漫长的道路上要以和为贵，和气才能生财；厅堂的天井有"四水归堂"之含义。按商人的心愿和希望，即代表着万象更新、四季发财的象征和愿望；厅堂太师壁两侧的走廊过道上各有一个"商"字造型。封建社会提倡"士农工商"，视商人为末等公民，引起了商人的强烈不满，为求得心理上的平衡，故在"过廊"上搞一"商"字造型，以表示自己高高在上，任何达官贵人从前厅到后堂都必须在"商"字的造型下走过；厅堂的柱子上悬挂着"传家之道"、"修身之道"、"经商之道"三副楹联。其中一副宣扬经商之道的楹联是：快乐每从辛苦得，便宜多自吃亏来。辛字多了一横，快字少了一竖，亏字多了一点，多字少了一点，这绝非是错别字，而充分表达了商人的思想境界。即：只有多吃点苦，才能多得到一份快乐；只有多吃点亏，才能多占一点便宜；也就是说，吃点苦不要紧，只要能赚钱；要赚钱就要吃点小亏，不要斤斤计较。

Ruiyu Hall–Lucky Jade Hall

Xidi Village, Yixian County

瑞玉庭窗栏板 Carved window boards

Built in the third year of the Xianfeng reign of the Qing Dynasty (1853), Ruiyu Hall is located in Xidi Village, Yixian County. On both sides of the door-dressing of the courtyard, there are two shoe-shape-gold-ingot-like brick carvings of two Chinese characters of "富" and "贵", meaning the blessing of bringing wealth and riches, rank and honor. On the right wall of the courtyard is the inscription of four Chinese characters—"履道含和", which means that one should be good-natured in his life, for amiability begets riches. The skywell of the lobby has a meaning of "receiving wealth". The two passages beside the screen wall of the main hall are of the designs of the Chinese character of "商"—business or businessman. The feudalist China advocated the social status order of "officials, peasants, craftsmen and businessmen", so the businessmen was the lowest in the society. Of course they were unsatisfied with that. To get their equilibrium, they designed this kind of structure to show their superiority, because any high officials had to pass under this design to get to the rear lobby during their visit. There are three vertical couplets hung on the columns of the hall—namely, "Teaching of the family heritage", "Approach of self-cultivation" and "way of doing business". In the couplet to blaze forth the way of doing business "快乐每从辛苦得，便宜多自吃亏来", people may find that a horizontal stroke was added to the character of "辛"; a point was added to "亏" and "多"; and a vertical stroke was deducted from "快". They were not written by mistake but were specially designed to reflect the motto of the businessmen—more hardships bring more happiness, and more sufferings invite more profits. That is to say, it does not matter to bear hardships for making money; one has to bear a small loss in order to make more money and he should not haggle over every ounce.

瑞玉庭大门入口 The entryway

瑞玉庭正堂 The main hall

瑞玉庭

西园

黟县西递村

西园位于黟县西递村横路街，建于清道光四年，即公元1824年，原为四品官、河南开封知府胡文照（号星阁）故居。大门是砖砌的八字门楼，显得气度非凡，进入八字砖雕门楼的大门，一字形并列摆开3个独立的三开间单元，由门前庭院相连贯通，古朴典雅的庭院居室便展现在眼前，在西递村众多的古民居中显得风貌特别，独具一格。

庭院虽以墙相分隔成前园、中园、后园，但是用青砖与大理石砌成的长方形大漏窗，相连通的圆月形、秋叶形、八边形门洞，使得整个西园庭院的景致均处在"隔与不隔，界与未界"之间。从大门门楼下望去，层次极为分明，景致颇称幽静。漫步观赏，透过漏窗，穿过门洞，随着观察点的不同，映入眼帘的均是一幅赏心悦目的图画。

更值得一提的是"黟县青"大理石的使用，使得"西园"又增添了无穷的艺术魅力。庭院内有石几、石凳、石桌、石井，还有石雕的门罩，里院的"松石"、"竹梅"石雕漏窗，构图清新，刀法细腻，镂空八层，更叫观赏者流连忘返。

West Garden

Henglu Street of Xidi Village

West Garden is in Henglu Street of Xidi Village, Yixian County. Built in the fourth year of the Daoguang reign in the Qing Dynasty (1824), it is the former residence of Hu Wenzhao (whose assumed name is Xing Ge), a fourth rank official of the court and prefecture governor of Kaifeng, Henan. The entrance is a splay gate tower made of bricks impressive in bearing. Behind the gate is three independent sections of three-bay houses in a row connected by the front yard. The simple and elegant courtyard residence is special in style and features among the ancient folk houses in Xidi Village.

The courtyard is divided into the front yard, the middle one and the rear one by walls, but the big oblong lattice windows made of black bricks and marbles, together with the moon gate, the leaf-shape gate, and the octagonal gate, make the three yards independent but not separated. The scenery is tranquil and well arranged, viewed from the upstairs of the gate tower. Through the lattice windows and the gates, the views are different but pleasing from various perspectives.

Moreover, the use of marbles of "Yixian County Black" adds more artistic charm to West Garden. There are benches, stools, tables, wells and gate shielding, which are all made of stone. The lattice windows of "Pines on Rocks" and "Bamboo and Plum" in the inner yard are delicate and pretty in structure. The hollowed-out eight plies are elaborately wrought and attract everyone.

西园

西园入口八字门　The 八-shaped door at the entrance

中国徽派建筑 古民居

西园前园院落 The front courtyard

西园中园院落 The middle courtyard

西园后园院落 The back courtyard

西园前园正堂 The main hall in the front courtyard

西园

西园院门 Gate of the garden

西园"松石"石雕漏花窗
The stone carved fretted window (Pine on Rocks)

西园门上匾额
The inscribed board above the gate

西园"竹梅"石雕漏花窗
The stone carved fretted window (Bamboo and plum)

西园存"西递"村牌
The name board of Xidi Village

西园"纶音锡命"匾额 The inscribed board

东园 \ 黟县西递村

东园位于黟县西递村，建于清雍正二年，即公元1724年，原为胡文照的祖居，是一组具有园林情趣，多单元的大型古住宅。

来到东园门前，主人胡星阁自题的行书"东园"门额眉刻赫然入目。眉刻左边有跋："颜以东园，志古也。古昔街之西名西园，柳下其东则曰东园。今人见此屋以居均忘乎其所自矣，故着而存云，俾访古者一览焉。屋后有井名东园井，是其一证云。星阁氏跋。"额如书画手卷，用"黟县青"镌成，风格典雅别致。

"东园"的厅堂结构颇为奇特，两边厢房的门呈相向对开，正对大厅。有趣的是其木雕图案、装饰造形却极不对称。左边厢房房门呈六边形，雕刻了"五福捧寿"组图，四只展翅的蝙蝠（同"福"音），意喻长寿。右边厢房房门则是菱形，是一幅很大的冰裂图，冰块棱角分明，冰棱寓意"十年寒窗苦"。

最为珍贵的是正厅天井墙上，嵌着的一块碑刻。此碑宽近70厘米，高约30厘米，从右往左刻有行草："结自得趣"四字，落款为"星阁四兄嘱题，鸿寿。"下方还有一回文"陈鸿寿印"章。

East Garden

Xidi Village, Yixian County

East Garden is in Xidi Village, Yixian County. Built in the second year of the Yongzheng reign of the Qing period (1724), it is the clan residence of Hu Wenzhao, which is a large-scale garden-like compound of ancient houses.

Above the gate, there is a horizontal inscribed board bearing the name of "East Garden" written by the owner Hu Xingge in running script, and on its left is the owner's notes of the construction. The board is made of "Yixian black stone", looking like a hand scroll in an elegant and novel style.

The structure of the halls is quite special—the doors of two wing rooms face each other and are opposite to the main hall. It is interesting that the designs of wood carvings and the decorations are asymmetrical. The door of the left wing room is hexagonal, inscribed with a series of designs of "Five Bats". Four flying bats (sounding the same as Chinese "囗") stand for longevity. The door of the right wing room is rhombus, which is a picture of rent ice. The edges and corners of the ice-pieces are distinct and clear with the meaning of "the hardships of ten year's study".

It is rare that a tablet was embedded in the wall of the skywell of the main hall. The tablet is 70 centimeters wide and 30 centimeters high. From right to left is the running script of "Jie Zi De Qu", with an inscription and a seal.

东园

东园入口 The entrance

东园正堂 The main hall

东园右厢房 The right-wing room

东园左厢房 The left-wing room

东园坐堂观 "福"
The carved character of "Happiness" on the wall

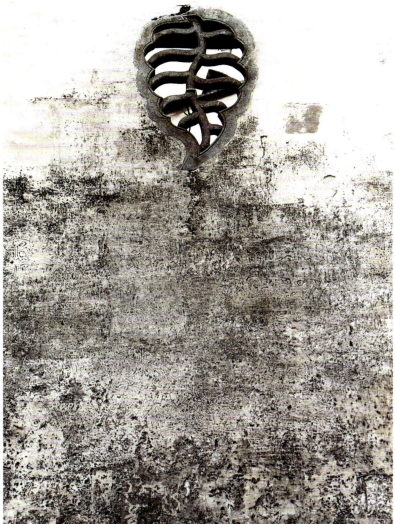
东园 "叶落归根" 砖雕 The leave-shaped brick carving

东园 "结自得趣" 碑刻 Carved tablet

履福堂 \ 黟县西递村

履福堂位于黟县西递村"司城第"弄内。建于清康熙年间，为收藏家笔啸轩主人胡积堂故居。"履福堂"分前后厅，三间三楼结构，高大宽敞，雅秀古朴。前厅挂有一幅很大的"松鹤"中堂画，中堂上方悬挂"履福堂"匾额，字体遒劲。中堂两侧和柱子上，挂有泥金木制楹联，上刻"世事让三分天宽地阔，心田存一点子种孙耕"，"第一等好事只是读书，几百年人家无非积善""诗书朝夕，学问性天，慈孝后先，人伦乐地。"等古训。在厅堂的长条案上摆设东瓶西镜（谐音平静，即平安吉庆之意）。中间放有自鸣钟（钟和终为谐音。古人云：在世一生无奢求，只愿终身有平静）。八仙桌置放文房四宝，厅两旁摆有罗汉椅。板壁挂有古代画家的字画，后厅堂前挂有一木质挂扇，一边刻着"清风徐来"四个大字，一边刻着"凌云"二字，一扯绳子就清风阵阵，多么惬意舒适。后厅天井设有一米高的金鱼池，摆设假山盆景，既能养鱼观赏，调节湿度，又有消防储水的功能，这在自古素有"旱船"之称的西递村确有不可低估的作用。楼上厅张挂有胡氏祖容。整座宅居古风盎然，书香扑鼻，具有中国古代典型的书香门第风貌。

Lüfu Hall–Hall of Experiencing Happiness

"Sicheng Di" of Xidi Village, Yixian County

Lüfu Hall is in "Sicheng Di" of Xidi Village, Yixian County. Built in the Kangxi reign of the Qing Dynasty, it is the former residence of Hu Jitang, a collector and owner of a veranda. The hall is divided into two parts, spacious and elegant with the structure of three-bay and three-story. A large scroll of "Pines and Cranes" is hung in middle of the front hall. Over the scroll is a plaque bearing the name of the hall. On either side of the scroll and the columns are hung the wooden couplets of old maxims in golden paint. On the long narrow table in the hall there are a vase on the east and a mirror on the west (sounding like "safe and quiet" in Chinese with the meaning of safety and luck). Right in the middle is a chime clock (sounding similar to "forever" in Chinese. The ancient saying goes, "Wishing placidity through one's life and no more extra demands.") The four treasures of the study are placed on the old fashioned square table. On both sides of the hall are placed arahant chairs. On the wooden partitions are hung the calligraphies and paintings of the ancient painters. In the front of the rear hall is hung a large wooden fan. On onee side of the fan is inscribed with four characters of "Breeze blowing slowly", and the other side is "Reaching the sky". It is pleasing and comfortable to have a cool breeze blowing when the string is pulled. In the skywell there is a pond for golden fish, about one meter high with rockery and bonsai, which could be used not only for the ornamental fish rearing and humility control, but also for the water storage for fire protection. This function should not be underestimated in Xidi Village which has been known as a "land boat" from ancient times. In the hall upstairs are hung the portraits of the ancestors of the Hus. The whole residence is overflowing with ancienrty, and the scent of ink is in the air. It has a typical style and features of literary families in ancient China.

履福堂正堂 The main hall

履福堂香案 The incense burner table

履福堂古挂扇 An old hanging fan

膺福堂 \ 黟县西递村

膺福堂位于黟县西递村,建于清代康熙三年(公元1664年),乃明经胡氏二十五世祖,诰封从二品,户部尚书胡如川(尚赠)的故居,现为其后裔居住。《膺福堂》高大贴墙门楼,飞檐翘角,方柱月梁,砖雕精美,气势宏伟,显示了非同一般的官第形制。大门入内置有仪门,每当主人家婚嫁喜庆或有身份的达官显贵光临才启门迎入。据说,七品官以下的一般人只能从仪门两侧的边门出入。整个大厅宽敞高大,既堂皇又肃穆,由上下厅堂、三间房和左右两庑两间厢房组成四合格局建筑,横贯大厅的东梁围达180厘米。枋斗雕镂花草禽兽,生动传神。大厅为礼仪往来的场所。大厅西侧有书房、鱼池厅和一座三间账房。后面是两进五间楼房。方厅柱、梁、枋均为方体,含方正廉明之意。方厅上有楼房,后进是三间楼房,往后是厨房、过道。中单元与东单元之间有一条贯通前后的深巷。东大门内是外东厅,装镶着24扇精工组制的花格门,厅前有一天井。这里是女眷和孩童习书的地方。里东厅是三间楼房,后进是厨房。整座民宅脊顶纵横,马头墙参差生姿,宏伟壮观,檐柱斜撑雕成倒吊的狮子,天井两边的厢房全用木雕花隔扇连接楼上楼下房间,古朴华丽,典雅别致,故有"翰林院"之称。

Yingfu Hall--Hall of Receiving Blessing

Xidi Village, Yixian County

Yingfu Hall is in Xidi Village, Yixian County. Built in the third year of the Kangxi reign in the Qing Dynasty (1664), it is the former residence of Hu Ruchuan (also named Shangzeng), the descendent of the 25th generation of the Hu Clan, a second rank official by imperial mandate and minister of the Ministry of Revenue, and now his descendants live there. The huge gate tower, adherent to the wall, has upturned eaves and ledges, square columns and crescent beams, and fine brick carvings. The imposing style shows its extraordinary official status. Inside the entrance is the ritual gate. The door would open only when there were ceremonies, or when high officials or noble lords visited the family. It was said that officials under the seventh rank could only enter the house through the side doors beside the ritual gate. The hall is high and spacious, serious and magnificent, in a closed arrangement including the up and down halls, three rooms, two corridors and two wing rooms. The east beam across the main hall is as long as 180 centimeters. The inscriptions of plants and beasts are lifelike and vivid on the architraves and dougong. The main hall is the place for gathering and ceremonies. On its west side are the study, the fish pond and the three-bayed accountant's office. Behind it are two rows of five-bay buildings. The columns, girders and architraves are all square in the square hall, which refers to honesty and uprightness. The square hall is a storied building and the rear row is a three-bay storied building, behind which are the kitchen and the aisle. The middle unit and the east unit are separated by a run-through long lane. Inside the east gate is the outer east hall furnished with 24 exquisite lattice doors. And in front of the hall is a skywell where women and children studied. The inner east hall is a building of three-bay and the rear row is the kitchen. It is imposing to see so many ridges and corbie gables here and there. The diagonal braces of the peripheral columns are inscribed into an upside down lion. The partition doors of wood carvings connect the rooms upstairs with rooms downstairs of the wing rooms beside the skywell. It is flamboyant and elegant, ancient and extraordinary; so it is also called "The Imperial Academy".

膺福堂香案 The incense burner table

膺福堂正堂 The main hall

笃敬堂

黟县西递村

笃敬堂位于黟县西递村，建于清康熙四十三年，即公元1703年，距今已有290余年，原为胡贯三之孙，胡如川之子胡积堂居住，现仍为其后裔居住，厅堂上最为醒目的便是胡氏二十六世祖胡积堂的遗像，从他的顶戴"蓝宝石顶"和补服"胸绣"孔雀……识别，应为正三品文官，遗容的两边堂柱上挂有一副内涵丰富，很有哲理的楹联："读书好、营商好、效好便好；创业难、守业难、知难不难。"联文将营商和读书并列提论，充分表达了徽商对提高自身地位的渴求和企盼。上联的"效"字是指学好和做好、读书、营商、两者一好，当然什么都好了；下联的"知"字是指认识和正视，只要我们认识困难，正视困难，就会迎难而上，化难为易。

笃敬堂窗栏木雕 Wood carvings on the window panels

Dujing Hall–Hall of Deep Respect

Xidi Village, Yixian County

Built in the forty-third year of the Kangxi reign of the Qing Dynasty (1703, about 290 years ago), it is located in Xidi Village, Yixian County. It is the former residence of Hu Jitang, the grandson of Hu Guansan and son of Hu Ruchuan, and now it is residence for his descendants. The portrait of Hu Jitang, the descendent of the 26th generation of the Hu Clan, is the most striking in the hall. Identified from his official cap (with a sapphire top) and gown (with an embroidery of peacock on the chest), he must be a third rank official in the court. On both sides of the portrait are a pair of meaningful and philosophic couplets, which equate business with study. This mirrors the demand and desire of the businessmen in Huizhou to raise their social status. The word "效--xiao" (efficiency) in the first line of the couplet indicates efficient study and business—all would be good if these two things are well done. The word "知--zhi"(knowing) in the second line means recognition and envisagement—as long as we recognize the hardship and face it, we would go upstream undauntedly and make everything easier.

笃敬堂入口 The entrance

笃敬堂门罩砖雕 The brick carved shielding above the gate

笃敬堂

笃敬堂天井 Skywell

笃敬堂前院 The front courtyard

大夫第

黟县西递村

大夫第位于黟县西递村正街。建于1691年，为朝议大夫知府胡文照故居。四合院二楼结构，正厅高大轩敞，厅前设天井。砖雕门罩上砖刻"大夫第"三个大字，正厅裙板隔扇均精雕冰梅图案，槛棂窗花仿明代格调。厅左侧利用隙地建有临街彩楼，俗称小姐绣楼，飞檐翘角，挂落、栏杆、排窗宽敞，玲珑典雅。楼额木刻隶书"桃花源里人家"，为清代黟县书法家汪师道所书；木刻小额"山市"二字，为清进士祝世禄手笔。楼下边门有石刻隶书"作后一步想"门额。

大夫第匾额 The inscribed board

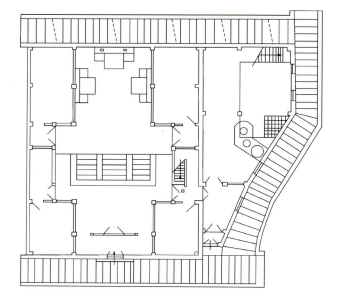

大夫第底层平面图 Plan of the first floor

Dafu Di–Senior Official Residence

Xidi Village, Yixian County

Dafu Di is in Zhengjie Street in Xidi Village, Yixian County. Built in 1691, it is the former residence of Hu Wenzhao, a prefecture governor at the time. The structure of the residence is a two-story courtyard. The main hall is spacious and bright with a skywell in front. On the brick carving of the gate shielding are the characters of "Da Fu Di". The panels and partition boards in the main hall are engraved with the design of plums in winter, and the window lattices are made to resemble the Ming style. On the left of the hall the opening was used to set up a decorated building, called "the lady's chamber". It has upturned eaves and petals as well as hanging fasciae and railings. The windows are large and bright, little and elegant. The wood carving on the head of the building is "Tao Hua Yuan Li Ren Jia—Families in the land of peach and plum" in official script by the calligrapher Wang Shidao of the Qing Dynasty. The small board is carved two words of "Shan Shi—Mountain Market" by Zhu Shilu, a Jinshi in the Qing period. The head jamb of the side door downstairs is the stone carving of "Zuo Hou Yi Bu Xiang—Give in a little and think it over" in official script.

大夫第入口 The entrance

中国徽派建筑　古民居

大夫第正堂　The main hall

大夫第

中国徽派建筑　古民居

大夫第天井 Skywell

大夫第观景楼　The viewing tower

笃谊堂 / 黟县西递村

Duyi Hall–Hall of Deep Friendship

Sanjiling Lane of Xidi Village, Yixian County

笃谊堂位于黟县西递村三级岭弄。又称"枕石小筑",建于清道光年间。三间二楼结构,宣统年间在右侧增建小筑。大门为砖砌八字门楼,上嵌砖刻"紫气东来"四字,大门里向有石雕"枕石小筑"横额,西侧建有瓶式、秋叶式小门洞,上有砖刻"玉壶"、"莺春"门额,庭院宽大,有石几、石凳、花木假山设置。正厅及左右次间的天花彩绘非常精美,斜撑、雀替描金飞彩,"小筑"的裙板隔扇木雕华丽。

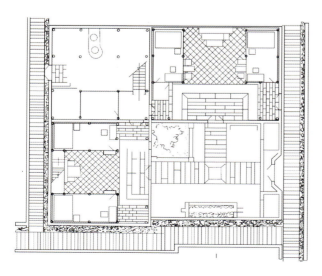

笃谊堂底层平面图 Plan of the first floor

笃谊堂入口 The entrance

笃谊堂

Duyi Hall is in Sanjiling Lane of Xidi Village, Yixian County. It was also named "Stone Pillow Building" and built in the Daoguang reign of the Qing Dynasty. It is the three-bay two-story structure and a little building was attached to it in the Xuantong reign. The entrance is the brick splay gate tower embedded with a brick carving of "Zi Qi Dong Lai". On the back of the gate is a horizontal stone board carved with "Zhen Shi Xiao Zhu". On the west are the vase-shape and leaf-shape small doors with the head jambs of brick carvings "Yu Hu--Jade Pot" and "Ying Chun—Early Spring". The yard is spacious with stone benches, stools, plants and rockery. The colored drawings on the ceiling are gorgeous in the main hall and the left and right subcentral halls. The diagonal braces and queti are painted in gold and color. The wood carvings are magnificent on the panels and partition boards.

中国徽派建筑　古民居

笃谊堂正堂　The main hall

笃谊堂天井　Skywell

笃谊堂

笃谊堂顶棚彩绘 Colour paintings on the roof

笃谊堂墙板彩绘 Colour paintings on the wooden wall

笃谊堂窗扇彩绘 Colour paintings on the window

笃谊堂顶棚彩绘 Colour paintings on the roof

承志堂 / 黟县宏村

承志堂位于黟县宏村。建于1855年前后，为清末盐商汪定贵住宅。砖木结构，全屋有木柱136根，大小天井九个，七处二层楼，大小60间，门60个，占地面积2100平方米，建筑面积3000平方米。全屋分外院、内院、前堂、后堂、东厢、西厢、书房厅、鱼塘厅、厨房、马厩等。还有搓麻将的排山阁，吸鸦片烟的吞云轩，以及保镖房、女佣住室、憩厅、小书房等。其平面布局具有典型的徽州民居特点，前院、福堂、寿堂位于主轴线上，纵向延伸，层层递进；东厢、西厢位于左右，对称布置；附属建筑顺应环境，因地制宜，形成一些不规则的平面和空间，各个空间都能巧妙地利用地形，创造出舒适典雅，别有洞天情趣。承志堂外院的大门，尺度不大，与路有一个角度，与风水有关，进入外院拾阶而上，面对的是一高大的八字门，旁边数米之外还有一小柴门，进入八字大门，便进入了幽深静谧、封闭豪华的内院空间。院内有长廊把各厅房联为一体，屋内有池塘、水井，用水都不用出屋。前堂是回廊三间结构，分上下厅，雕梁画栋，天井四周为锡打水枧，上有"天锡纯嘏"四个大字，在仪门的上方雕有一幅"百子闹元宵"图和一个斗大的"福"字，"百子闹元宵"雕刻一百个小孩，有的打锣，有的敲鼓，有的放鞭炮，有的吹喇叭，有的踩高跷，有的划旱船，真可谓五花八门，样样都有。活灵活现地表现了徽州民间闹元宵的喜庆场面。两边的侧门的"商"字门头，又好似两个古币，色彩和雕刻都十分精美，内容雕有"三英战吕布"、"战长沙"、"长坂坡"等三国故事，喻意含蓄而深刻。福厅的额枋上刻有一幅"唐肃宗宴官图"，图中四张八仙桌一字排开，众官员琴棋书画各得其乐，

Chengzhi Hall—Hall of Behest-Inheriting

Hongcun Village, Yixian County

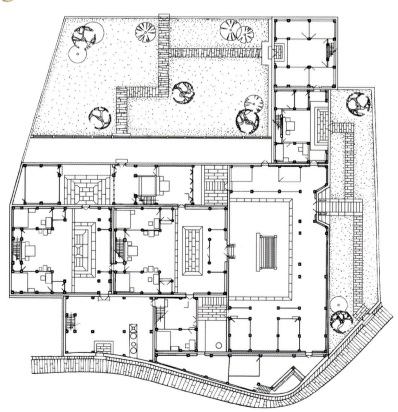

承志堂底层平面　Plan of the first floor

饮酒猜拳乐此不疲，坐站行止姿态各异，左右两端还有剃头、烧水之辈，这一切人物形态逼真，呼之欲出。额枋后的卷棚造型独特，呈"Ω"型，两端各有一队金狮滚绣球，绣球镂空雕，四只金狮倒挂金钩背朝下，观后令人叹服。福厅两边厢房的双扇莲花门亦是令人赞叹不已，门扇上部为镂空雕，雕有暗八仙图，门芯和门板雕有渔樵耕读、福禄寿喜、吉庆有余图案，人物描金着色，显得格外富丽堂皇。福厅的柱础均为"元宝"础，寓意财源广进。后堂和前堂的结构基本相同，内院亦有轿廊，用以停放桥子。不同的是寿厅是长辈居住的地方，故所有的雕饰都与"孝悌忠信"有关。额枋上雕有"郭子仪上寿图"，图中三位老人端坐高堂，有一对晚辈跪在堂前，双手合十，给老人拜寿，两边的文官武将神态各异，对晚辈的孝心频频称是。寿厅南边的梁枋上雕有"百忍图"，告诫晚辈孝敬长辈和治家要以忍为先。柱础均四方柱础，四面都雕"寿"。寓意长辈们寿比南山。在轿廊的西侧是鱼塘厅，呈三角形结构。据传当时建造承志堂，设计时靠村中水圳，多了这块三角形空地，工匠便设计了这个三角形建空间。鱼塘厅小天井下有一汪鱼池，圳水从屋外潺潺流进，又通过石栏栅向前流去。池畔有"美人靠"，可凭栏观鱼。多余数十平方米建成小客厅和小居室，鱼池的对面有一幅"喜鹊登梅"石雕，四只喜鹊鸣、跳、飞、栖姿态各异，枝枝梅花争芳斗艳。不仅雕刻精巧，层次分明，线条清晰，而且构图巧妙，不愧为石雕精品。另据传当时建造承志堂花白银60万两，其中木雕上镀黄金100两，全屋所有木雕由20个工匠雕刻四年才完成，"承志堂"可谓是徽州民居的极品和代表作。

Chengzhi Hall is in Hongcun Village, Yixian County. Built in 1855 around, it was the residence of Wang Dinggui, a businessman of salt at the end of the Qing period. It is the brick-and-timber construction and there are 136 pillars, 9 courtyards and skywells, 7 two-story buildings, 60 rooms and 60 doors. The hall extends 2100 square meters and the constructions cover an area of 3000 square meters. The hall includes the outer compound, inner compound, front hall, rear hall, east and west wing rooms, studies, fish ponds, kitchens and stables. Besides, there are a hall for mah-jong playing and a hall for opium-taking as well as the rooms for bodyguards and maidservants, and small studies. Its plan arrangement has typical features of the folk hoses in Huizhou—the front yard, the Hall of Happiness, and the Hall of Longevity stand one after another on the central axis; the east wing room and the west wing room are on the sides symmetrically; the attached constructions make up irregular closed spaces according to the circumstance; every closure makes use of the land forms to create a unique elegant and dainty world. The entrance of the outer compound is not very big and forms an angle with

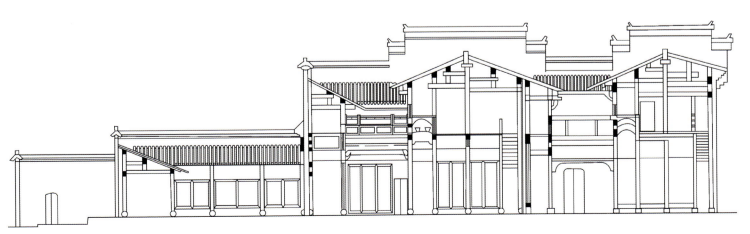

承志堂剖面 Section view

承志堂一进福厅 The Hall of Happiness in the first section

the road according to Fengshui. Inside the outer compound there is a high splay gate, and a small side door is a few meters away from it. Behind the splay gate is the deep and splendid inner compound. The porches there connect all the rooms and halls into a whole. It is unnecessary to go out to fetch water because there are ponds and wells inside. The front hall is the three-bay construction with cloisters, including the up and down halls, with carved beams and painted rafters. Around the skywell are the tin pipes. Above the ritual door is a large carved character of "Fu—Happiness" and the design of "A Hundred Children Celebrating the Lantern Festival" in which there are one hundred children—some are beating gongs and drums; some are letting off firecrackers; some are trumping; some are stilting; some are rowing land boats. The wide variety of entertainments vividly reflect the celebrations people held on the Lantern Festival. The □-shape door-heads of the side doors look like two ancient coins with refined colors and carved stories from the Three Kingdoms. On the architrave of the Hall of Happiness is the inscription of "A court banquet offered by a Tang emperor" in four old fashioned square tables

stand in a row, and the officials enjoy themselves in various postures and expressions, every figure being vividly portrayed. The shape of the round ridge roof behind the architrave is out of the common in "Ω" shape, with a design of golden lions sporting with the silk ball on both ends. It is admirable that the silk ball is hollowed up and four lions are hanging upside down. The two lotus doors of the wing rooms in the Hall of Happiness are really beyond all praise. The upper part is the through-carved work with the design of the hidden Eight Immortals. On the door panels and boards are the engravings of designs of "Fishermen, Woodcutter, Peasants, and Intellectuals", "Fortune, Emolument, Longevity and Happiness" and "Fishes for Luck". It is magnificent that the figures are gilded and colored. The column plinths in the Hall of Happiness are of the ingot-shape with the meaning of profits pouring in from all sides. The structures of the rear hall and front hall are quite similar, with a porch for sedan-chair. But, because the Hall of Longevity is the residence for the elders, all the decorations and inscriptions here are related to filial pieties. On the architrave is the carving of the design of "Guo Ziyi's

承志堂二进寿厅 The Hall of Longevity in the second section

Birthday Celebration". The three elders in the design are sitting up in the hall, and two juniors are kneeling in front and offering birthday felicitations with palms together devoutly. The civil officials and military officers beside rain praises on the juniors in different manners. On the south beam architrave is the inscription of "A Hundred Forbearance" which intends to preach to the juniors to take tolerance and forbearance as the first consideration in treating the elders and in managing a household. The plinths are square with carvings of character "□—longevity" on four sides to show the blessing of long life of the elders. On the west of the sedan-chair porch is the Hall of Fish Pond in the structure of triangle. It was said that this triangle place was left over and used for this construction when the residence had been designed to build backing the ditch in the village. There is a pond in the skywell in the Hall of Fish Pond and the ditch water flows from outside through the stone fences. Beside is beauty's chairs for sitting and watching the fish. The extra area of about tens of square meters is used for a small living room and a small bedroom. Opposite to the pond is a stone carving of "Magpies on the Plums Tree"—

announcing good news, in which four magpies are singing, jumping, flying and perching and the plum blossoms contend in beauty and fascination. It is really an elaborate stone-carving because it is exquisite in sculpture, distinct in gradation, clear in lines and peculiar in composition. As it was said, 600,000 liang of silver was spent on building Chengzhi Hall and 100 liang of gold was used on the wood-carings. It took 20 craftsmen four years to finish all the wood-carvings in the residence. Chengzhi Hall is really the best and the representative work of the folk houses in Huizhou.

承志堂八字门楼 The 八-shaped entrance

承志堂柴门 The side door

承志堂入口 The entrance

承志堂后院 The back courtyard

承志堂彩绘顶棚 The colour-painted roof

承志堂鱼塘厅 Fish Pond Hall

承志堂堂前船篷轩顶棚 The round ridge roof

承志堂天井水枧 The water pipe of the skywell

承志堂莲花门 Lotus doors

承志堂喜鹊梅石雕窗 Stone carved window of "Magpies on the Plum Trees"

承志堂正门倒座 The reverse side of the entrance

承志堂二进寿厅 The Hall of Longevity in the second section

承志堂"商"字门头 The 商-shaped door head

承志堂"百子闹元宵"木雕 Wood carving of the Lantern Festival

承志堂"唐肃宗宴官图"木雕 Wood carving of a court banquet

承志堂"郭子仪上寿图"木雕 Wood carving of a birthday celebration

承志堂"百忍图"木雕 Wood carving of "one hundred of forbearance"

承志堂

承志堂斜撑　The bracket

承志堂斜撑　The bracket

碧园

黟县宏村

碧园位于黟县宏村牛肠水圳头附近，明末始建，后毁，重建于清道光十五年（公元1825年），占地面积278平方米，建筑面积256平方米。该宅院建筑朴实洗练、幽静雅淡。院门向南，门口为一泉活水。楼房却坐东朝西，三间二楼形式，楼厅窗扇规整明快，可开可卸，登高远眺，山峦平野尽收眼底。高瞰西溪，自然临水，景致变化莫测，却把宅院同外景拉近了。

绝妙的是庭院水榭设置别出一格。你看庭内沿水圳傍石立基紧靠楼房正西掘有鱼塘，楼门口塘沿设有美人靠栏杆，走出厅堂即入水榭，如襟带环绕，楼间隐榭，水际安亭，实在灵巧别致。观鱼赏花各听其便，使人胸襟舒坦，醒心惬意。塘西设置石长条摆设各式花盆，有乔木数株，花坛数座，漏窗数幅，疏淡的花影，翠绿的枝叶，把鱼池也染绿了。

塘东沿有花墙一幅，弯道曲处有通，实处有虚。院北侧新建楼房一幢，其中有鸟瞰全村的鹤寿亭，有卵石铺地休闲养心的花果园，新老建筑互为补充，亭楼水榭集于一园。令人赏心悦目、心旷神怡。

碧园是宏村清代庭院水榭民居的代表作之一。

Bi Yuan – The Green Garden

Hongcun Village of Yixian County

碧园庭院 The courtyard

The Green Garden is located near Niuchang ditch in Hongcun Village of Yixian County. It was first built at the end of the Ming Dynasty and was destroyed later. Rebuilt in the fifteenth year of the Daoguang reign of the Qing Dynasty (1825), it covers an area of 278 square meters and the constructions occupy 256 square meters. The residence is plain and simple, elegant and quiet. The gate faces south and there is an active fountain in front of it. But the buildings face west in the three-bay two-story structure. The movable windows are neat and luminous. Looking far into the distance, you can have a panoramic view of the mountains and plains. Looking down at the Xixi Brook, the scenery is changing constantly with the water. Thus, the residence and the scenery outside become close to each other.

It is excellent that the arrangement of houses and yards adopted an original approach. In the yard stone bases are set up along the ditch, and on the west of the building there is a fish pond. In front of the building there are beauty chairs along the pond. The ribbon-like water surrounds the houses. It is novel that the pavilions are set up beside the water and hidden among the buildings. You can either watch the fish or enjoy the beauty of flowers. Long narrow stone strips are placed to the west of the pond to display various potted flowers and trees and rockery.

There is a lattice wall and a snaky path to the east of the pond. On the north of the yard is a newly-built building. By the building is a pavilion where you can take a bird's-eye view of the village, and a cobblestone-paved garden where you can rest and repose. It is pleasing and joyous that the new and old buildings complement each other. The pavilions, buildings and the water make a perfect harmony in the garden. The Green Garden is one of the representative works of the Qing folk houses with waterside pavilions in Hongcun Village.

碧园水榭 Waterside pavilion

碧园美人靠 The beauty chairs

八大家 \ 黟县关麓村

八大家位于黟县关麓村。始建于清乾隆年间,是一户汪姓徽商八个兄弟的住宅。八大家相对独立为八个单元自成一体,各有天井、厅堂、花园、小院,但又互相通联,屋楼上下皆有门户将其串结,形成一个整体。八大家的大堂描金绘彩,木雕精美,虽历数百年,仍极鲜妍。户户均有清一色的铁皮大门和莲花小门。

八大家把古建筑的套联艺术与室内装饰艺术融为一体,独领风骚。

据《徽州地区简志》记载:迁徽氏族多以自己的始祖或迁祖为中心,集居繁衍,形成宗族,一般一族居一村,有的累世同居。关麓"八大家"就是典型的代表,它是清朝汪氏徽商昭公所生八子及后裔聚族而居、相继构筑的豪华联体宅第。

八大家占地约1公顷,共有大小民居近30幢。其分布,由南向北,与县内清代著名书画家汪曙故居"武亭山房"相接。"武亭山房"建于清康熙年间,现尚存半壁门楼。边邻为一排三幢廊步三间,建于清咸丰年间,乃"八家"老三家后裔居住。向西为"涵远楼",俗称"学堂厅",系回廊三间,后有藏书楼,建于清同治年间,乃"八家"老五令镳所建。

继续往北,经一幢廊步三间,为"吾爱吾庐"横向连片建筑,乃"八家"老大令銮于清咸丰年间建造,其子德浩早年读书于此。"吾爱吾庐"为书斋式建筑,回廊三间,其门额系晚清著名书法家赵之谦题书。

隔石板路往北,又是一排三幢廊步三间,乃"八家"老四令钰在清同治年间所建。其中西向一户"大夫第"堂前,摆着一张200多岁的抽烟床,长约2米,宽1米余,系用珍贵的"海底木"制成,与其配套的还有1个中央茶几和2个踏脚架等,无不精雕细镂,令人叫绝。沿小溪北行,为"八大家"最早的民居建筑"春满庭"建于清乾隆年间,系回廊大型四合屋,乃"八家"老八家后裔居住,亦为黟县著名抗日烈士汪希直的故居。庭前有大院墙,经围墙内门通往另一幢廊步三间"瑞霭庭",为"八家"老四令钰,又名征三建于清同治年间。后有回廊相连,通向"双桂书屋"。西向隔石板小巷为一四合屋,乃"八家"老四令钰之子德沆于民国初年所建。

再隔一石板路往北,有带小庭院的三间屋,乃"八家"老六令钟之子德澄于民国二年所建。西向胡同内为"安雅书屋",曲径通幽。隔壁为三幢横排相连的廊步三间,乃"八家"老二令铎于清咸丰年间建造,内有"淡月山房"。往北为前后两进廊步三间和旁置便厅的廊步三间各一幢,均乃"八家"老七家于清同治年间建造。

沿小溪再往北,石板桥前、八字门楼为前后两进廊步三间,乃"八家"老六令钟于清同治年间所建。前进大门上方有"大夫第"三字门额,后进有便厅、"临溪书屋"等。隔壁是一幢三间屋,为清咸丰年间所建。屋内有一大月洞门，通向"八家"最著名的经学堂"问渠书屋"，昔日书屋内建有亭台楼阁，水榭回廊等，惜今已毁坏。

庞大的"八大家"民居群，各家既相对独立、自成一体、各有厅堂、天井和花园、庭院等，但又互相通达，屋楼上下有门户将其串结，共成一个整体，结构别致，设计精巧。外人进家，如入迷宫。

"八大家"家家是白墙青瓦小窗户，门楼门罩墙壁画。还有那参差错落、形似马头的"封火墙"，俗称"马头墙"，在青山绿水中煞是醒目、气派。"八大家"户户都有清一色的铁皮大门和莲花小门。在相连两家的过道处，门户极为狭窄，但过道中央却有宽敞的便厅。据说便厅内的八仙桌只好在里面打就，以后就永搬不走了。

自古人们都认为房屋坐北朝南，才能冬暖夏凉，而"八大家"大门却大都朝北，而不朝南。这是因为汉代流行着"商家门不宜南向，征家门不宜北向"的说法。

"八大家"古民居的砖、木、石"三雕"艺术非常精美。门楼、墙门、花窗等外墙面镶有各种砖雕饰品，增添了三维效果。大小木雕艺术作品在厅堂居室内纷呈显现。其花纹图案大都追求高雅和吉祥。幢幢民居的大堂前更是雕梁画栋，富丽雅致，照壁两侧边门的上方，大都用4幅木雕构图拼成大大的"商"字。石雕主要施于勾栏踏步、柱础、门框、漏窗等部件，其力度不凡，动势不俗，别有韵味。

"八大家"厅堂天花、居室内壁的彩绘艺术亦堪称一绝。大都根据建筑物的功用及居住对象的不同而绘以和谐鲜明的水彩画面，虽历经百年，今仍鲜妍。天花是我国传统建筑中顶棚天花上的一种装饰处理。"八大家"的天花彩绘非常独特，其形状多为民间喜闻乐见的吉祥物，并由外框图案和框内画面两部分组成。如用蝙蝠图案作为外框，框内配画一个"寿仙"；用双钱作为外框，里面一画"寿仙"，一画"子孙满堂"，象征"福寿双全"（因黟方言，"钱"与"全"谐音）。有的外框画一石榴，框内画一个胖娃娃或是桔子，寓意"来（榴）生贵子"；有的画如意或钟、鼓，象征"万事如意"、"晨钟暮鼓"之意；还有的为体现"清白传家"，便画一棵大白菜；也有的画各式各样的鱼，寓意"年年有余"，并特别以画青、白、鲤、鳜四种鱼居多，取其谐音即为"清、白、礼、贵"。这些寓意深刻的天花图案，形态逼真，栩栩如生。

居室内板壁窗门上的彩绘，堪称美术一绝。主要内容有"麒麟送子"、"富贵（牡丹）花开"、"五子登科"、"孔融让梨"及"冰清玉洁"（即画一冰裂纹花瓶内插一枝梅花图案）等，还有其他山水风景、人物故事等图案，无不形象生动，呼之欲出。这些彩绘艺术，既装饰美化了居住环境，陶冶性情，又能融传统的道德教育于耳濡目染之中，确乎费了一番心机。而且，这些彩色壁绘，对于今天的美术爱好者，也有极高的鉴赏和临摹价值。

Residence of Eight Prominent Families

Guanlu Village, Yixian County

The residence of eight prominent families is located in Guanlu Village, Yixian County. First built in the Qianong reign of the Qing period, they were the residence of eight brothers of the Wangs, who were businessmen at the time in Huizhou. The houses of eight families are eight separate units with skywells, halls, gardens and yards, but they are connected with each other and there are doors in each house leading to other units upstairs and downstairs, thus joining them as a whole. The main hall is decorated in gild and color and the wood carvings are delicate. Though they have existed for hundreds of years, they are still very vivid and bright. Each family has the gate of sheet iron and the small lotus-like door.

The design of these houses combined the art of interlink and the art of interior decoration in harmony, which leads the fashion of the time.

According to Brief Record of Huizhou District, the clans, which had migrated to Huizhou, usually lived around their first settler or the ancestor leading the migration and then multiplied there. Thus, a clan came into existence and usually lived in one village. Some of them had been living there for many generations. The "Eight Prominent Families" is a typical representative and it is the luxury compound which the eight sons of Wang Zhaogong, a businessman in the Qing period, and their descendants built in succession and lived in.

The residence occupies an area of about one hectare and has about 30 houses, big or small. They are arranged from south to north and connected with Wuting Villa where Wang Shu, a famous painter and calligrapher in the Qing Dynasty, used to live. Wuting Villa was built in the Kangxi reign in the Qing period and now only half of the gate tower exists. Beside it is a row of three three-bayed houses with corridors, which were built in the Xianfeng reign of the Qing Dynasty. The third of the eight brothers and his descendants lived here. To the west is Hanyuan Hall, also called "School Hall" by the natives. It is three-bayed with cloisters, and a library is behind, which was built in the Tongzhi reign by the fifth brother Lingbiao.

Further on the north through a three-bay corridor are the cross continuous constructions called "I Love My Cottage", which were built by the eldest brother Lingluan in the Xianfeng reign. His son Dehao used to study here when he was young. "I Love My Cottage" is a study-style architecture with three-bay cloisters. The inscribed board above the door was written by Zhao Zhiqian, a well-known calligrapher in the late Qing Dynasty.

Northward across the stone-slab road there is another row of three three-bayed houses with corridors, which were built in the Tongzhi reign by the fourth brother Lingyu. In the "Dafu Di", facing west, is a 200-year-old smoking bed which measures about two meters in length and more than one meter in width. It is made of precious "Undersea Wood" and accompanied by a teapoy and two footboards. Everything is delicately carved. On the north along the brook is the clan's oldest house of "Chun Man Ting" built in the early Qing period. It is a large-scale courtyard house with cloisters and the eighth brother and his descendants have lived there. It is also the former residence of Wang Xizhi, a famous anti-Japanese martyr in Yixian County. There are large bounding walls in front of the hall. Its inside door leads to another three-bay house with corridors named "Rui'ai Ting", which was built in the Tongzhi reign of the Qing period by the fourth brother Lingyu who was also named Zhengsan. Behind it a cloister leads to "Twin Laurels Study". Across the slab lane there is a courtyard house facing west, which was built by Dehang, the son of Lingyu, at the beginning of the Republic of China.

North further across a stone-slab road are three houses with a small yard, which were built by Decheng, the son of the sixth brother Lingzhong, in the second year in the Republic of China. In the west lane there is "Anya Study" which is quiet and deep. Next to it are three continuous three-bayed houses with corridors, built in the Xianfeng reign by the second brother Lingduo. Inside it is "Dan Yue Villa". On its north are a two-depth three-bay house with corridors and a three-bay house with an informal hall, which were built by the seventh brother in the Tongzhi reign.

Northward along the brook, in front of the slate bridge are the splay gate tower and two sections of three-bay house with corridors, which were built by the sixth brother Lingzhong in the Tongzhi reign. Above the gate of the first section is an inscribed board of "Dafu Di". In the second section there are an informal hall and a study. Next to it is a three-bay house built in the Xianfeng reign. Inside the house, there is a big moon gate leading to the most famous school of Confucian classics of the clan—"Wen Qu Study". The pavilions, terraces, towers, and cloisters had been set up there but were destroyed later.

The colossal compound of structures has separate and independent units with halls, skywells, gardens and yards of their own. But all the units are linked up with each other. Doors upstairs and downstairs connect each other into a whole. The arrangement is novel and the design is unique. Strangers would have the feeling as if he went into a maze.

All the houses have the whitewashed walls, gray tiles and small windows as well as the gate towers, gate shieldings and frescos. There are irregular horse-head-shape fire-sealing corbie gables, called "Horse-head gable" in local name. They are striking among the blue hills and green waters. Each house has the gates of iron sheet and small lotus doors. The doors of the aisles between the two connected houses are very small, but in the middle of the aisles, there are spacious informal halls. It is said that the old-fashioned square table has to stand there and cannot be moved out of the hall.

Since ancient times, people have thought that houses should face south to have warm winters and cool summers. But all the houses here face north because of the saying in the Han Dynasty—"The doors of businessmen should not face south and those of warriors should not face north".

The carvings of bricks, wood and stones are exquisite and splendid in the residence of "Eight Prominent Families". The external side of the gate towers, wall doors, and lattice windows are decorated with various brick carvings showing a three-dimensional effect. The art works of woodcarvings are displayed in the halls and rooms, and the designs are graceful and propitious. The main halls are richly ornamented with carved beams and painted rafters. Above the side doors of the screen walls there is usually the character of "shang" assembled by four wood carvings. The vigorous and novel stone carvings are adopted in the railings, steps, plinths, doorframes and lattice windows, which have a lasting appealing.

The colored drawings on the caissons and the inner walls can be rated as superb works of art. The bright and harmonious watercolor paintings of various patterns were drawn according to the different functions of the rooms and the different dwellers. Though they have existed for hundreds of years, they are still bright and vivid. The caisson is a kind of decoration on the ceiling in the traditional Chinese architectures. The colored drawings on the caissons in this residence are peculiar and the patterns are usually the propitious symbols popular with common people. The drawings consist of a frame pattern and a picture pattern. For instance, if the bat design is used in the frame, the picture pattern must be the god of longevity; if the design of two coins is adopted as the frame, there must be a drawing of the god of longevity and a drawing of many sons and grandsons, meaning completeness in both respects of felicity and longevity (in the dialect of Yixian, "钱" and "全"have the similar sounds). All these patterns on the caissons are meaningful and vivid.

The colored drawings on the inner walls are also the superb works in art. The contents are mostly "Kylin bestowing a baby-child", "Peonies of wealth and rank", and "Pure and Noble" (a twig of plum in a vase with the pattern of ice crackles). Besides, there are landscapes and stories of people. Everything is lifelike and vivid. The ancient people really beat their brains out to make these colored drawings. They not only furnish the living circumstance and exert a favorable influence, but also function traditional moral instruction in daily touch. These murals are worth appreciating and facsimileing even to today's art-lovers.

八大家全貌 A full view of the Residence of the Eight Prominent Families

八大家

"吾爱吾庐"—老大家　House of the eldest brother—"I Love My Cottage"

"春满庭"—老八家　Courtyard in the eighth brother's house

"双桂书屋"—老四家 Study in the fourth brother's house

中国徽派建筑　古民居

"瑞霭庭"—老四家正堂　The main hall of the fourth brother's house

老八家彩绘　Colour paintings in the house of the eighth brother's

"敦睦庭"一隅—老六家 A corner of the hall in the sixth brother's house

老八家彩绘 Colour paintings in the house of the eighth brother's

"吾爱吾庐"—老大家树叶门 A leave-shaped door

老六家前院 Front courtyard in the sixth brother's house

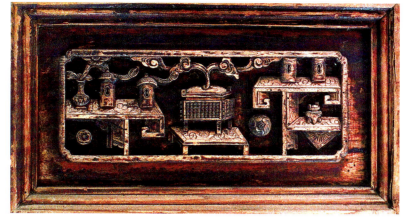

八大家绦环板 Carved panels

迎祥庭

黟县关麓村

Yingxiang Hall

Guanlu Village, Yixian County

迎祥庭位于黟县关麓村，建于清末同治年间。迎祥庭的入口很有特点，虽然在路旁，却没有对着路开门，而是在路边留出一个小广场，对着广场两边开了一个正门和一个边门。正门上有门楼，石门框中嵌有木雕花挂落，两侧有八字墙，仿八字门楼。边门十分简洁，拱形门上匾额，书"迎祥"二字。穿过正门是前院，一条石板路通往宅门，院落一侧的围墙正对宅门开有一长方形漏花窗。进入宅门是一进天井，正堂三开间，两侧是厢房。整个房屋的雕刻都以木本色为主，给人一种朴实清新的感受。

迎祥庭天井 Skywell

迎祥庭

Built in the Tongzhi reign at the end of the Qing Dynasty, Yingxiang Hall is in Guanlu Village, Yixian County. The entrance is peculiar—though the hall is situated by the road, its main gate and the side door are on the two sides of a small square between the road and the hall, instead of opening towards the road. Over the entrance is a gate tower and in the stone door frame is embedded a wood carving of festoon. There are splay walls on either side of the entrance, imitating the splay gate tower. The side door is very simple with a plaque of "Ying Xiang" over the arch door. Behind the main gate is the front yard, and there is a stone-slab road leading to the door of the house, which is opposite to a rectangle lattice window on the enclosing walls. There is a skywell inside the door of the house and the main hall is the three-bay construction with two wing rooms. It is simple but delicate and pretty that all the wood carvings in the houses have the natural color of the wood.

迎祥庭入口　The entryway

迎祥庭正堂 The main hall

迎祥庭

冰凌阁

黟县南屏村

冰凌阁位于黟县南屏村，建于清初康熙年间。由于场地限制，入口退路边1.5米，做成八字门楼，水磨砖砌筑，背北朝南。进门是四方的院落，有一条青石板小路通往正厅和偏厅，其他地面铺有鹅卵石。院落的左侧是正厅，二层三开间，正堂悬挂着"怀德堂"的匾额，二层依然保存着昔日祭祀用的香火座和焚香炉，另外在两侧的厢房都有一高大的壁橱。院落的右侧是回廊，回廊极具特点。有一个圆形的木雕棋门，雕刻有冰梅和花鸟，十分精美，整个回廊仅1米多宽，却给整个庭院起了锦上添花的作用。进门正对的是偏厅，偏厅亦为二层，一层莲花门上方悬挂着一块"冰凌阁"的匾额，莲花门上半部镶嵌玻璃，玻璃上绘有梅、兰、竹、菊等图案，十扇莲花门的板芯则雕刻有"西湖十景"。偏厅的二楼，中间是厅，两侧是厢房，靠天井一侧是一条走廊，外向是"美人靠"，走廊两端各有一个2米高的拱形门洞，一端是采光通风之用，另一端装有一扇门，开门可通往正厅的二层。倚立在二层的"美人靠"旁，远处的青山绿水和院落中的情景一览无余。

Bingling (Icicle) Hall

Nanping Village, Yixian County

Built in the Kangxi reign at the beginning of the Qing Dynasty, Bingling Hall is in Nanping Village, Yixian County. Facing south and limited by the space, the entrance withdraws 1.5 meters back from the road and is made of terrazzo in a splay gate tower. Inside the gate is a square courtyard and a lane of bluestones leads to the main hall and the side hall, the rest of the yard ground being paved with cobbles. On the left of the yard is the two-story three-bay main hall and inside the hall is hung a plaque of "Huai De Hall". On the second floor there are the sacrificial incense burner and seats used in old days. In either of the wing rooms, there is a tall closet. On the right of the yard is a peculiar cloister. The circle archway of wood carving is delicately sculpted with icy plums and birds and flowers. The whole cloister is only one meter wide but adds brilliance to splendor. The door is opposite to the two-story side hall. A plaque of "Bingling Hall" is hung above its lotus door. Glass is embedded in the upper part of the lotus door with the patterns of plums, orchids, bamboos and chrysanthemums. The panels of the ten lotus doors are inscribed with "Ten Scenic Spots of the West Lake". In the middle of the second floor is the hall with wing rooms on both sides. Along the skywell is a corridor with "beauty chairs" at the outer side. At the ends of the corridor are the arch doorways of two meters high. One is for daylighting and aeration and the other is installed with a door leaf open to the second floor of the main hall. Sitting on the beauty chair upstairs you can have a full sight of the courtyard nearby and the faraway blue mountains and green waters.

冰凌阁入口 The entrance

冰凌阁正堂 The main hall

冰凌阁木圆洞门 The wooden round door

冰凌阁偏厅　The side hall

木雕楼

黟县卢村

木雕楼位于黟县卢村，建于清代道光年间，系奉政大夫、朝议大夫卢帮燮古居。木雕楼入口与众不同，从路边拐进一个小棋门，在你面前呈现是一个狭长的前院，一条半米宽的石板路贯穿两端。前院的中部有一门楼，是一个回廊的入口，穿过回廊，即是志诚堂的前院，院落两侧有围墙，且对称的开有一对漏花窗，一条1.5米宽的石板路直对大门，巨大的牌楼式门罩显得格外气派。进门是一进天井，三开间，两侧厢房，二层对天井一侧均设有莲花门，且在天井的四周设有一圈连通的回廊。回廊宽不足50厘米，外侧有漏空花的栏杆。此宅最大的特色是无论楼上、楼下、栏板、窗板、门扇、墙裙、梁柱、雀替、斗栱、檐口等都有令人叹为观止的精美雕刻。那"陶翁醉酒"、"苏武牧羊"、"姜太公钓鱼"无不雕刻得栩栩如生。那"舞龙灯"、"垒建新屋"、"集市繁荣"无不雕刻得活灵活现。站在这木雕的世界中，细细地品味每一幅作品，你将不由得对徽派文化的博大精深和古代工匠精湛的技艺肃然起敬。

木雕—商贾繁荣
Wood carving of "prosperous market"

Wood-carving Building

Lucun Village, Yixian County

Woodcarving Building is in Lucun Village, Yixian County. Built in the Daoguang reign of the Qing period, it is the former residence of Lu Bangxie, a senior official in charge of the consultation and court discussion. Its entrance is out of the ordinary. Behind the small arch door on the roadside is a long and narrow front yard, through which is a stone-slab path of about half a meter wide. In the middle of the front yard is a gate tower which is the entrance of a cloister. At the end of the cloister is the courtyard of Zhicheng Hall. On both sides of the yard are bounding walls and there are two symmetrical lattice windows in the walls. A slab road about 1.5 meters wide leads to the main gate. The huge archway-style gate-shielding is really imposing. Inside the gate is a three-bay construction with a skywell and two wing rooms. There are lotus doors upstairs facing the skywell and a connecting cloister around the skywell. The cloister is less than 50 centimeters wide with fretted railings at the outer side. The admirable characteristic of this residence is the exquisite sculptures on the railing panels, windows, door leaves, wall panels, beams, columns, queti, dougong, and eaves upstairs and downstairs. Everything is vividly and lively inscribed. When enjoying in this world of sculptures, one cannot help admiring the extensive and profound culture in Huizhou and respecting the exquisite skills of ancient craftsmen with deep esteem.

木雕楼前院及入口 Front courtyard and the entrance

木雕楼外院 The external courtyard

中国徽派建筑 古民居

木雕楼正堂(志诚堂) The main hall (Zhicheng Hall)

中国徽派建筑　古民居

木雕楼二层　The second floor

木雕楼前天井　The front skywell

木雕楼后天井　The back skywell

木雕楼

木雕楼前厅堂侧面　Side of the front hall

二层屋檐一角　Corner of eaves on the second floor

木雕—垒建新屋　Wood carving of "building a new house"

木雕—陶翁醉酒　Wood carving of "a drunk old man"

再版后记

中国徽派建筑文化展开幕式嘉宾，左起：
赵晨、高帆、郑孝燮、干志坚、张开济、罗哲文、周谊、杨永生、刘慈慰、樊炎冰

杨永生、罗哲文、傅熹年先生在展会上

本书作者樊炎冰先在开幕式上讲话

当我校对完最后一页书稿时，心潮澎湃，眼前又浮现出2001年，《中国徽派建筑》大型画册在北京举行首发式时的情形，原建设部长汪光焘；中国摄影家协会主席高帆先生；中国文物学会会长罗哲文先生；中国历史文化名城保护委员会主任郑孝燮先生；建筑大师张开济先生；建设部副部长干志坚先生等专家和前辈均出席了开幕式，令我感动至今。十年后的今天，我们在原《中国徽派建筑》画册基础上，重新编印的《中国徽派建筑》付梓之际，更加深深怀念已经离开我们的高帆先生、张开济先生、干志坚先生、彭守仁先生.深深感谢他们在中国古建筑保护与研究上所做出的杰出贡献。

我也深深感谢带我走入研究徽州古建筑之门的张文起先生，还有原徽州区旅游局局长许智勇先生，我在考察徽州建筑的十年间先后数十次到访古徽州，是他带我跑遍了古徽州的一府六县，考察了数百个有价值的徽州古建筑，为本书的积累了大量文字和图片资料。我也要深深感谢安徽省、市、县各级领导和工作人员，他们为协助我考察提供了无私的帮助。我也要感谢东南大学的陈薇教授和龚恺教授，他们无偿的为本书提供了大量的徽州建筑测绘图。感谢中国建筑工业出版社张振光先生，他为本书的编辑出版付出了辛勤的劳动，最后也要感谢我太太和女儿，在后方给了我巨大的帮助和支持。

新版的《中国徽派建筑》一书，是在原《中国徽派建筑》的基础上，做了大量的修改和补充，加上了英文翻译，使其更加国际化，版式和开本也做了调整，精简了一些项目，增补了一些有价值的新项目，在图片上也精益求精，相信本书的出版会给人耳目一新的印象，也相信本书会为中国徽州古建筑的研究提供更详实的内容和更精美的形象素材。

徽州古建筑博大精深，沉淀了中国千年文化，本人才疏学浅，十多年的不耻下问，潜心研究编成此书，旨在为保护祖国的文化遗产尽一份微薄之力。该书新版已对前书的谬误予以修正，但仍难免有所纰漏，恳请各位专家不吝赐教。

樊炎冰
2011.11于广州

Postscript for the Revised Edition

本书作者与中国摄影家协会主席高帆先生

罗哲文、陈铎、高帆等在展会上

中国徽派建筑文化展策展人合影，左起樊炎冰、周谊、陈铎、赵晨、刘慈慰、闫东

When I finished proofreading the last page of this revised edition, I was very excited. I couldn't help recalling ten years ago the opening ceremony of this large-size album - Huizhou-School Architecture of China in Beijing. It was my great honor to have Mr. Wang Guangtao, former minister of Ministry of Construction; Mr. Gao Fan, Chairman of China Photographers Association; Mr. Luo Zhewen, Chairman of China Heritage Academy; Mr. Zheng Xiaoxie, director of Chinese Historical Cultural City Protection Committee; Mr. Zhang Kaiji, master architect; Mr. Gan Zhijian, vice minister of Ministry of Construction, and many other specialists to attend the opening ceremony. Ten years after that, we've compiled a revised edition of the original Huizhou-School Architecture of China. When it is about to be published, Mr. Gao Fan, Mr. Zhang Kaiji, Mr. Gan Zhijian, and Mr. Peng Shouren who have left us are especially remembered. My deep gratitude will be given to them for the contribution they have made in Chinese traditional architecture protection and research.

I would like to thank Mr. Zhang Wenqi, who tutored me when I first started to study traditional Huizhou-School architecture. And Mr. Xu Zhiyong, former chief of Huizhou District Tourism Bureau, who took me to the one prefecture and six counties of ancient Huizhou during my dozens of visit to Huizhou in these ten years. The survey of hundreds of valuable Huizhou traditional architecture has accumulated a large amount of text and pictorial materials. Similarly, I'm grateful to the directors and working staff of Anhui Province, cities and counties, who have provided generous help for my survey. I would also like to express my gratitude to Professor Chen Wei and Professor Gong Kai of Southeast University, who unselfishly offered numerous survey drawings for this publication. My special thanks go to Mr. Zhang Zhenguang from China Architecture & Building Press for his consistent conscientious input. Finally, I would like to thank my wife and my daughter, who have always been there helping and supporting me.

Based upon the original edition of Huizhou-School Architecture of China, I've done a lot of changes and additions to the revised edition. And the English translations make it more international. Layout and format are also adjusted, a number of projects are deleted, while some valuable new projects are added; at the same time, the pictures are refined. I believe that this publication will give people a fresh impression, and will provide much more detailed content and vivid image of the ancient Chinese Hui-style architecture.

Extensive and profound ancient Hui-style architecture is the deposition of thousands of years of Chinese culture. I have to admit that I know very little in this field. It took me over ten years of research and learning from specialists to finally complete this book, which I've been devoting myself to protect the cultural heritage of my motherland. In this revised edition, I've tried my best to correct the mistakes of the former edition. But I know there is going to be certain flaws remained, please don't hesitate to leave comments and give kind instructions.

Fan Yanbing
Nov.2011, Guangzhou

附：主要参考书目

[明]弘治《徽州府志》，上海古籍书店1964年影印本。

[清]吴逸：《古歙山川图》，清刊本。

[民国]吴吉祐：《丰南志》，稿本。

歙县地方志编纂委员会：《歙县志》。

休宁县地方志编纂委员会：《休宁县志》。

婺源县地方志编纂委员会：《婺源县志》。

祁门县地方志编纂委员会：《祁门县志》。

[清]康熙《徽州府志》，清刊本。

[清]江忠俦、江正心：《新安景物约编》，清刊本。

张仲一等：《徽州明代住宅》，建筑工程出版社1957年版。

季家宏：《黄山旅游大辞典》中国科学技术出版社，1994年10月。

程极悦：《徽派古建筑》黄山书社。

刘敦桢：《刘敦桢文集》，中国建筑工业出版社，1987年。

同济大学城市规划教研室：《中国城市建设史》，中国建筑工业出版社1982年版。

梁思成：《营造法式注释》，中国建筑工业出版社1983年版。

刘敦桢：《中国古代建筑史》，中国建筑工业出版社1980年版。

叶显恩：《明清徽州农村社会与佃仆制》，安徽人民出版社1983年版。

中国建筑技术发展中心建筑历史研究所：《浙江民居》，中国建筑工业出版社1984年版。

梁思成：《梁思成文集》，中国建筑工业出版社1986年版。

李允鉌：《华夏意匠》，中国建筑工业出版社1985年版。

中国科学院自然科学史研究所：《中国古代建筑技术史》，科学出版社1985年版。

张海鹏、王廷元：《明清徽商资料选编》，黄山书社1985年版。

罗哲文：《中国古塔》，中国青年出版社1985年版。

茅以升：《中国古桥技术史》，北京出版社1986年版。

《中国建筑史》编写组：《中国建筑史》，中国建筑工业出版社1986年版。

山西省建筑保护研究所：《中国古建筑学术讲座文集》，中国展望出版社1986年版。

罗哲文：《中国古代建筑》，上海古籍出版社，1990年。

刘致平：《中国建筑类型及结构》，中国建筑工业出版社1987年版。

中国大百科全书出版社编辑部：《中国大百科全书——建筑 园林 城市规划》，中国大百科全书出版社1988年版。

汪之力等：《中国传统民居建筑》，山东科学技术出版社，1994年。

丁宏伟：《徽州明清祠堂建筑》，载于《建筑历史与理论研究文集》，中国建筑工业出版社1997年版。

单德启：《中国传统民居图说——徽州篇》，清华大学出版社1998年版。

中国民族建筑出版委员会：《中国民族建筑》第四卷。

陈薇：《江南包袱彩画考》，载于《建筑理论与创作》，东南大学出版社1988年版。

国家文物局：《中国名胜词典》上海辞书出版社，1986年。

何晓昕：《风水探源》，东南大学出版社1990年版。

中国建筑史编委会：《中国建筑史》第二版，中国建筑工业出版社1997年。

潘谷西：《我国明代地区中心城市的建筑》，载于《建筑历史与理论研究文集》，中国建筑工业出版社1997年版。

东南大学建筑系、歙县文物事业管理局：《徽州古建筑丛书》，东南大学出版社1996——1999年版。

高寿仙：《徽州文化》，辽宁教育出版社1998年版。

陈从周：《中国民居》，上海学林出版社，1993年。

图书在版编目（CIP）数据

中国徽派建筑 / 樊炎冰主编．— 北京：中国建筑工业出版社，2011.9
（世界文化遗产）
ISBN 978-7-112-13514-1

Ⅰ．①中… Ⅱ．①樊… Ⅲ．①古建筑-建筑艺术-徽州地区 Ⅳ．①TU-092.2

中国版本图书馆CIP数据核字（2011）第174222号

责任编辑：张振光　杜一鸣
英语翻译：张本慎
装帧设计：肖晋兴
责任校对：肖　剑　姜小莲

世界文化遗产
中国徽派建筑
樊炎冰　主编
＊
中国建筑工业出版社出版、发行（北京西郊百万庄）
各地新华书店、建筑书店经销
北京方嘉彩色印刷有限责任公司印刷
＊
开本：787×1092毫米　1/8　印张：48½　字数：605千字
2012年3月第一版　2012年3月第一次印刷
定价：480.00元
ISBN 978-7-112-13514-1
　　　（21253）

版权所有　翻印必究
如有印装质量问题，可寄本社退换
（邮政编码　100037）